Stamps Tell the Story of Space Travel

Also by Emery Kelen

Dag Hammarskjold: A Biography
Fifty Voices of the Twentieth Century
Hammarskjold: The Dangerous Man
Hammarskjold: The Political Man
Let's Learn About the United Nations
Peace in Their Time
Peace Is an Adventure
Proverbs of Many Nations
Stamps Tell the Story of John F. Kennedy
Stamps Tell the Story of the UN
Stamps Tell the Story of the Vatican
The Valley of Trust

Edited and annotated by Emery Kelen

Fantastic Tales, Strange Animals, Riddles, Jests, and Prophecies of Leonardo da Vinci

Stamps Tell the Story of Space Travel

by emery kelen

THOMAS NELSON INC.
Nashville New York

Acknowledgments

I am greatly indebted to Marvin Robinson and Wladimir V. Azarenkov, senior officers of the Outer Space Affairs Division of the United Nations, for their courtesy in reading the book.

I also wish to express my gratitude to J. & H. Stolow, Inc., 915 Broadway, and Enzo Buonocore Company, 829 Second Avenue, stamp dealers in New York, for having provided me with the stamps photographed in this book.

Published in Nashville, Tennessee, by Thomas Nelson Inc. and simultaneously in Don Mills, Ontario, by Thomas Nelson & Sons (Canada) Limited. Manufactured in the United States of America.

First edition

Library of Congress Catalog Card Number: 74–181677
International Standard Book Number: 0–8407–6212–7
0–8407–6213–5 NLB

Contents

Foreword *9*

Introduction *11*

Man's Ancient Dream

Copernicus *13*
Kepler's Dream *15*
"E Pur Si Muove" *17*
Sid Isaac Newton *19*
Jules Verne *20*
George Méliès *21*
Konstantin E. Tsiolkovsky *22*
Robert H. Goddard *24*
Rockets *26*

At the Beginning, There Was Sputnik. . . .

Sputnik I *28*
Lunik Series *30*
Laika, the Space Dog *31*
Explorer I *32*
Luna III *33*
Tiros Series *35*
Cosmos Satellites *36*
Flying Zoo, USSR *37*
Space Puppy *38*
Flying Zoo, USA *39*

Man in Space

Yuri Gagarin *41*
The Original Seven *42*
Alan B. Shepard, Jr. *43*

Virgil I. Grissom *44*
Gherman S. Titov *46*
President John F. Kennedy Looks into *Friendship 7* *47*
Rendezvous in Orbit *49*
First Woman in Space *50*
More Americans in Space *51*
Three Men in a Spacecraft *52*

SATELLITES EXPLORE SPACE

Scientific Satellites *54*
Syncom Satellites *56*
France's Satellites *57*
Canada's Alouette *58*
Lunar Exploration *59*
The Lunas and the Rangers *60*
Moon Rush *61*
Unmanned Flight Around the Moon *62*

SPACE WALKS AND OTHER EVA'S

Space Walk *64*
Gemini Series *65*
Extravehicular Activities (EVA's) *66*
Gemini IV *67*
Docking *70*

TRAGEDY STRIKES TWICE

Tragedy on the Launch Complex *71*
Death of a Cosmonaut *72*

THE APOLLO PROGRAM

Apollo 7 *74*
Meeting with Santa Claus *75*
Genesis *77*
Which Way to the Moon? *78*

Apollo 9 *80*
The Most Expensive Suit *81*
Apollo 10 *83*
Snoopy and *Charlie Brown* Hugging *85*

MEN ON THE MOON

Apollo 11 *87*
Columbia and *Eagle* *89*
". . . One Giant Leap for Mankind" *91*
The Moon Is Not Made of Green Cheese *92*
Moon Chores *94*
Moon Rocks *96*
Back to Earth *98*
Peace for All Mankind *99*
Splashdown! *101*

IS THERE LIFE ON THE MOON?

Biological Tests *103*
Lunokhod I *105*
Apollo 12 *107*

A SPACE ODYSSEY

Apollo 13 *109*
Abandon Ship! *111*

ANOTHER MOON EXPLORATION

Apollo 14 *113*
Football-Sized Moon Rocks *116*
Lunar Golf Balls *117*

PROBING THE SPACE AROUND US

Mission to Mars *119*
Luna XVI *120*
Mission to Venus *121*

Orbiting Shish Kebab *123*
"Bridge to the Future" *124*

ANOTHER SPACE TRAGEDY

The Hatch Was Opened—Three Men Were Dead *126*

THE SPACE AGE IS HERE TO STAY

Post Office on the Moon *128*
Space Technology—Its Practical Applications *130*
Medical Miracles Through Space Research *132*
International Cooperation *134*
The UN in Action *135*
A Last Word *137*

INDEX *139*

Foreword

There is perhaps no other field of human endeavor where science and technology have made such far-reaching progress in the past years as in the exploration and use of outer space. . . .

Scientific spacecraft traveled as far as the planets Mars and Venus; other spacecraft circled the moon and transmitted pictures of its hitherto unknown side. Spacecraft landed on the surface of the moon, gathering scientific information about the composition of the lunar surface. The first pioneers of space exploration have orbited the earth and, leaving their spaceships, have stepped out into outer space. At the same time, the practical applications of space research and technology are becoming, to an ever greater extent, part of our everyday life. . . .

The United Nations have not failed to recognize the important role they can play in this respect. In the first place it appeared imperative to prevent an extension of the armaments race into outer space. . . .

In the Treaty Governing the Activities of States in the Exploration and Use of Outer Space, Including the Moon and Other Celestial Bodies, . . . far-reaching provisions were established limiting military activities in outer space.

The United Nations have also concentrated their efforts on the establishment of legal rules for the exploration and use of outer space. As early as 1963 the General Assembly adopted a Declaration of Legal Principles Governing the Activities of States in the Exploration and Use of Outer Space.

The General Assembly . . . adopted a second treaty, the Agreement on the Rescue of Astronauts, the Return of Astronauts and the Return of Objects launched into Outer Space. . . .

The United Nations have thus taken the first steps to bring the rule of law into outer space. . . .

Outer-space research and its practical applications are, by their very nature, of global concern and will require to an ever-increasing degree the cooperation of all nations. . . .

It is the firm wish of all of us that the exploration and use of outer space should be carried out in the interest and for the benefit of all mankind, and that the fruits of this great human endeavor should be shared by all nations irrespective of the degree of their economic and technical development.

Dr. Kurt Waldheim,
Secretary General of the United Nations

Introduction

Man has only begun his exploration of outer space and has only started to implement the application of this new knowledge to his understanding of planet Earth. As with the scientists and engineers, the United Nations, too, is attempting to relate the accumulative experience of the international community to this new area of activity.

There is little question but that this new adventure of man will have tremendous influence on the scientific and technical growth of the future as well as upon the political, social, legal, and economic future of man. We in the United Nations who have, from the very outset, followed the exploration of outer space are aware that too many of our fellowmen have only the faintest indication of what the peaceful uses of outer space can mean to them in improving their lives here on Earth.

It is therefore our hope that this book will inspire the reader to delve further into the subject of man's peaceful use and exploration of outer space.

A. H. ABDEL-GHANI
Chief, Outer Space Affairs Division,
United Nations

Copernicus (1473–1543)

Man has probably always been awestruck by the celestial bodies—the changing shapes of the moon, the rising and setting of the sun, and the myriads of stars. Special astronomical phenomena, like the appearance of a comet or an eclipse of the sun or moon, filled him with dark forebodings.

Yet he was eager to go where the gods live and angels play their harps. He imagined winged lions, flying carpets, flaming chariots, and other fantastic means for flying about. In 1500 B.C., King Kai Ka'us of Persia (credited also with building the Tower of Babel) attempted to fly to heaven by fastening four eagles to his throne.

Icarus, the ancient Greek myth tells us, used wings of bird feathers, sticking them on his back with wax, but as he flew too near the sun the wax melted and poor Icarus plunged to his death.

The story goes that in eleventh-century Britain, a Benedictine monk named Eilmer fastened wings to his hands and feet and flew an eighth of a mile before crashing. He blamed his misfortune on "forgetting to put a tail on his back part" and for using wing feathers of chickens, which are known to be poor flying birds.

Among the Greek cosmologists, Plato (427?–347 B.C.) believed that the Earth was the center of the universe, with the moon, the sun, and the planets rotating around it. This concept was accepted by Aristotle (384–322 B.C.). The achievements of Greek astronomy were summed up by Ptolemy, who worked in Alexandria, Egypt, in the second century after Christ and postulated that the Earth was at the center and was immovable, with the moon, the sun, and the planets orbiting in perfect circles. The medieval world believed, with Ptolemy and Aristotle, that man and his Earth were the center of the universe and that the church was the heart of society on Earth.

Then came a timorous Polish astronomer, Nicolaus Copernicus, who contradicted this theory. Copernicus theorized that the planets orbited around the sun and not around the Earth, as it was previously believed. He developed his theories in a book that was not published until thirty-six years later. The first copy was given to him two hours before he died.

Copernicus was a great theoretician, but his immediate scientific successor in Denmark, Tycho Brahe (1546–1601), was too conservative to accept his theories. Fearing to offend the Catholic Church, he offered a compromise by proposing that all planets orbited the sun, but the sun orbited the Earth.

The Polish People's Republic in central Europe is about the size of New Mexico. Key industries are nationalized and operated under a planned economy. About 85 percent of the farms and some businesses are privately operated.

Population: 33 million. Area: 121,000 square miles. Capital: Warsaw.

Colors: slate blue and white

Kepler's Dream

Johannes Kepler (1571–1630), a German astronomer, became known to Galileo and Tycho Brahe, the leading astronomers of the time, when he published a book that upheld the teachings of Copernicus. Kepler became Tycho's assistant and later his successor. He is best remembered for his laws on planetary motions, which are based on Tycho's observations of the planets.

Kepler also wrote a book called *Somnium* (*"Dream"*) which, to avoid political and religious censure, he presented as a kind of science fiction. In it he tells of a fantastic journey from Earth to the moon with the help of demons. He describes the moon as inhabited by bizarre creatures who do not live in cities but hide in caves and crevasses and come out only to sun themselves for a short period. These fantasies served to disguise Kepler's scientific message, for he described in *Somnium* the gravity of the Earth, the dangers to Earthmen of rarified air, and the probable climate and landscape of the moon. Kepler credited his mother for his supposed supernatural transportation to the moon, and since the book was semi-autobiographical, his mother was jailed, and it took him five years of struggle against the authorities to prevent her from being burned at the stake by the Inquisition, which prosecuted heresy.

Israel was reestablished as a nation in 1948. It occupies part of the ancient land first called Canaan, and in modern times called Palestine.

Israel's economy and industrialization have both grown rapidly. Citrus fruit is the most valuable agricultural product. Israel's first atomic reactor began operation in 1960.

Population: 2.9 million. Area: 8,000 square miles. Capital: Jerusalem.

Colors: sky blue, white, gold on navy-blue background

"E Pur Si Muove"

In 1608 Galileo Galilei (1564–1642), the great Italian astronomer and physicist, heard that Hans Lippershey, a spectacle maker in Holland, had constructed a simple magnifying instrument by using two lenses held one in front of the other. Galileo, knowing no more than that, worked out the principle involved and proceeded to build the first complete astronomical telescope, which brought an object thirty times nearer. Through his homemade telescope Galileo looked at the celestial bodies and discovered that the moon is not smooth like marble, but pockmarked; the Milky Way is made up of numberless stars; Jupiter has four moons; Venus, like the moon, has phases; and that the sun has spots. Galileo eloquently defended the theories of Copernicus.

The system of Copernicus, which demoted the Earth to being just one of a number of planets, was denounced as dangerous to the Catholic faith, and Galileo was summoned to Rome and warned not to uphold the theory or teach it. But he persisted. Again summoned to Rome, he was tried by the Inquisition and condemned to abandon all beliefs that held the sun to be the central body and the Earth and other planets to be moving about it. It is said that when Galileo, kneeling before his judges, arose, he whispered, referring to the Earth: *"E pur si muove."* ("Nevertheless, it does move.") He died blind and under house arrest.

The stamp was issued by the Republic of Italy, which occupies a long peninsula, shaped like a boot and extending southeast from the Alps into the Mediterranean. Industry and modernized agriculture have enjoyed an extraordinary growth in recent decades. The lure of classical ruins, Renaissance art, imposing churches, and good living attracts more than thirty million tourists a year to Italy.

Population: 54 million. Area: 117,000 square miles. Capital: Rome.

Colors: violet, Indian red, cream

Sir Isaac Newton (1642–1727)

Sir Isaac Newton, an English mathematician and philosopher, is recognized as one of the greatest scientists who ever lived. He was farmer's son and, seeing that famous apple fall from a tree, started his study of gravity, which keeps man rooted to the ground. It also keeps the planets and their satellites exerting a mutual pull on each other. He set out the principles of dynamic astronomy and provided the central idea of the contemporary concept of the universe.

In his *Principia Mathematica,* Newton scientifically explained the basic principle of rocket propulsion. His second law of motion states that reaction is equal but opposite to action; or, the actions of two bodies are equal but point in opposite directions. In other words, for every action there is an equal and opposite reaction. If you see the tail of a rocket backfire, you will see the rocket will fly forward.

The Republic of Paraguay is one of the landlocked countries of South America, bound by Bolivia, Brazil, and Argentina. Most of the population is agricultural and pastoral. Cattle breeding is the principal industry.

Population: 2.4 million. Area: 157,000 square miles. Capital: Asunción.

Colors: royal purple, white, mandarin orange

Jules Verne (1828–1905)

The nineteenth-century science-fiction writers knew that eagle wings and balloons would not do the job of getting to the moon. They realized that for a trip to the moon much more power was needed to escape the Earth's gravity. The most amazing among these writers was the Frenchman Jules Verne. In his book *From the Earth to the Moon,* published in 1865, he described how three men sitting in a padded rocket were shot out of a cannon to start a lunar journey. He correctly calculated that a velocity of about 25,000 miles an hour would be required for a spacecraft to escape the Earth's gravity. With amazing foresight he even chose Florida as the takeoff site, and the Pacific Ocean for the splashdown. He described weightlessness and anticipated that a test flight of animals would precede a manned launching. Dogs, chickens, a cat, and a squirrel traveled to the moon in his story, and on the way the cat ate the squirrel. But Verne couldn't figure out how to bring his passengers back from the moon. So his rocket only orbited the satellite and landed safely in the Pacific.

Jules Verne's influence on rocketeering was enormous. Hermann Oberth (1894–), the German physicist and mathematician who was one of the early pioneers of rocket development for interplanetary flight, said that as a young man he read Verne's book six times and learned it by heart.

The Republic of France borders on Belgium, Luxembourg, Germany, Switzerland, Italy, Andorra, and Spain. It is separated from England by the English Channel and the Strait of Dover.

Population: 51 million. Area: 211,000 square miles. Capital: Paris.

Color: blackish green-blue on white

George Méliès (1861–1938)

The first fictional proposal for a manned space station came from a thirty-seven-year-old Bostonian, Edward Everett Hale, who wrote the story of "The Brick Moon," a satellite placed into polar orbit to serve as a navigational aid for sailors.

In 1897, H. G. Wells, an English historian and science-fiction writer, described an invasion of Earth by Martians in *The War of the Worlds.* Only their vulnerability to Earthly bacterial infection saved our world from being conquered.

The movie industry, still in its infancy, was inspired by the idea of space travel. The French movie pioneer George Méliès produced a film, *Journey to the Moon,* in which scientists were shot to the moon by a rocket.

To celebrate the centenary of George Méliès's birth, the French postal service issued this stamp.

Agriculturally, France is a country of small farms; 15 percent of the employed work on 46 million acres. France is the biggest food producer in western Europe and the most important vinery in the world. The country is rich in minerals. Manufacturing includes chemicals, textiles, perfumes, automobiles, aircraft, ships, and electrical equipment.

Colors: dark turquoise, dark violet, cinnamon

Konstantin E. Tsiolkovsky (1857–1935)

This Russian scientist was one of the first to understand the use of rockets in space travel. Tsiolkovsky arrived at a theoretical solution of how to escape the Earth's atmosphere and gravitational field. In 1903 he prepared the design of a spaceship to be powered by liquid oxygen and liquid hydrogen, a very modern combination of propellants.

The Russian schoolteacher also discovered the principle of the multistage rocket, which he called a "rocket train." He was convinced it was the only means by which a space vehicle could attain the velocity needed to leave the Earth's gravitational field. As each state used up its propellant fuel, it would be discarded to keep the weight of the vehicle down while the next stage would take over the job. Tsiolkovsky foresaw that rockets could visit planets or their satellites and return to Earth. "The Earth is the cradle of the mind," he wrote in 1896, "but one cannot live forever in the cradle."

The Union of Soviet Socialist Republics (USSR) is, in area, the largest country in the world. It stretches across two continents—Europe and Asia—and from the North Pacific to the Baltic Sea. There are some 150,000 rivers and 250,000 lakes in the country. The larger European rivers include the Dnieper, Volga, Ural, and the Don. The USSR is incalculably rich in natural resources. Atomic power stations have been built, and others are under construction.

Population: 242 million. Area: 8,647,000 square miles. Capital: Moscow.

Colors: ultramarine blue, pale blue, light and dark sepia

Robert H. Goddard (1882–1945)

Goddard, the American rocket pioneer, was first attracted to space flight by reading Verne's *From the Earth to the Moon.* In a 1904 high school oration, Goddard said, "It is difficult to say what is impossible, for the dream of yesterday is the hope of today and the reality of tomorrow."

His college research convinced him that rocket propulsion was the key to space travel. He believed that liquid oxygen and liquid hydrogen were the best fuels for rocketeering. An illustration in a scientific paper he published in 1919 with the Smithsonian Institution showed Goddard's concept of a rocket that could reach the moon.

On March 16, 1926, Goddard's liquid-fuel rocket made the first such flight in world history. It flew 184 feet in 2.5 seconds. The American press took a dim view of the theories of this scientist from Massachusetts, whose ideas ran counter to the accepted scientific notion that an engine could not operate in a vacuum.

In 1929, after launching a rocket that made an infernal noise and created a sea of flames, he was prohibited from further experimenting by the Massachusetts authorities. So he moved to New Mexico, where his rocket flights reached 9,000 feet.

When Sputnik I was launched on October 4, 1957, space

exploration came of age. But Goddard was already dead. The United States government bought his 214 patents from his widow for one million dollars.

In 1502 a map of the New World showed a promontory on the east coast of Florida. That desolate area was swampland and barren stretches of sand; for hundreds of years it remained a hunter's and fisherman's paradise. In 1845 Florida became part of the United States.

America's space age was ushered in at this spot. In 1950 the Air Force converted Cape Canaveral into a missile base. Now called Cape Kennedy, it is the most important spaceport in the United States, consisting of twenty-five miles of launch pads, radar domes, liquid-oxygen fuel plants, and the world's second-largest structure: the Vehicle Assembly Building, 125 million cubic feet of air-conditioned space. A visitor's information center handles some three million tourists who visit Cape Kennedy each year.

Colors: smoky blue, vermilion, sandy, off-white

Rockets

Rocket propulsion is the basis of all space flight. No other known form of man-made locomotion can impart the thrust necessary to boost an object to a speed overcoming the gravitational pull of the Earth.

You can observe muscle-powered jet propulsion by watching a squid, a member of the mollusk family. By siphoning water into its body and then ejecting it, the squid is able to move forward, a living example of Newton's law of action and reaction. This was the principle the Chinese, in A.D. 1040, applied to their rockets, which were propelled by black powder, an explosive mixture of saltpeter, charcoal, and sulfur.

In 1241 the Tartars used rockets against the Poles. In 1249 the Moors used them in their campaigns against the Spaniards. In 1420 an Italian military engineer designed a variety of military rockets—some of them looked like pigeons, fish, and rabbits. In 1650 Kazimierz Siemienowicz, a Polish artillery expert, summed up the existing knowledge of rocketry in *Artis Magnae Artilleriae.*

Rocketry changed the concept of warmaking. By the end of World War Two, the Germans were launching thousands of V-2 rockets against England. The first long-range rocket used, this 46-foot-long, 14-ton guided missile had a range of over 200 miles and attained velocities of up to 3,500 miles per hour. The Soviet Union's 6-foot-long *Katyusha* rockets had a distance of three miles. The British were firing their 10.5-foot-long, 740-pound missile *Stooge,* which had a velocity of 500 miles an hour. By July, 1944, the United States had produced about one million 134-pound aircraft rockets, which they called *Holy Moses.* Other rockets bore the whimsical names of *Tiny Tim, Little Joe, Bat,* and *Gargoyle.*

The Hungarian People's Republic in central Europe is bounded by Czechoslovakia, the USSR, Rumania, Yugoslavia, and Austria. The Danube River bisects the country. The eastern half is a great fertile plain, the west and north are hilly. Industries include machines, chemicals, and railways; agriculture's most important products are grains, vegetables, and the world-famous Tokay wine.

Population: 10.3 million. Area: 36,000 square miles. Capital: Budapest.

Colors: marine blue, tan, white

AT THE BEGINNING,
THERE WAS SPUTNIK. . . .

Sputnik I

By June 24, 1952, the United States National Advisory Committee for Aeronautics resolved that it should ". . . devote a modest effort to problems associated with unmanned and manned flight. . . ." Then, on October 4, 1957, the Russians orbited the first man-made satellite of the Earth, Sputnik I.

The news was made public at a cocktail party given in the Russian Embassy in Washington in honor of the delegates attending the meeting of a special committee for the International Geophysical Year.

Sputnik I (the word means "fellow traveler") was a very modest satellite compared to later satellites. It was an aluminum alloy sphere 22.8 inches in diameter, weighing only 184 pounds. It was equipped with two radio transmitters and four antennas, which made it look like a floating mine going *beep-beep*.

Sputnik I remained in orbit, gradually losing altitude, until January 4, 1958, finally disintegrating as it reentered the denser regions of the atmosphere. But its faint *beep-beep* was loud enough to put Washington on the alert. In April of that year, President Eisenhower sent a formal message to Congress to establish a civilian space agency.

Although some saw a disastrous blow to American prestige in the launching of Sputnik I, others belittled the Russian achievement by calling Sputnik I a "silly bubble in the sky" that "would be forgotten in six months." But on November 3, 1957, the Soviet Union launched another "silly bubble," Sputnik II. This sphere weighed 1,120 pounds and carried a dog named Laika. Russian achievements in space exploration were now firmly established and acknowledged.

The German Democratic Republic (East Germany) is bordered by the Baltic Sea in the north, by Poland in the east, by Czechoslovakia in the southeast, and by the Federal Republic of Germany in the southwest and west. The northern part of the country is lowland, and the elevation increases gradually as one travels south. Forests cover the southern mountain ranges.

Population: 16 million. Area: 42,000 square miles. Capital: East Berlin.

Colors: dark gray and white

Lunik Series

The first three Soviet moon probes were called Luniks I, II, and III by the American press. The Russians called them Cosmic Rockets I, II, and III, until they decided to name their moon probes Lunas.

The first of the Luna series, Luna I, was launched by the Russians on January 2, 1959, but it missed the moon. Eight months later, the Russians hit the moon with Luna II and demonstrated that they had more powerful rockets than the Americans did at that time.

The Mongolian People's Republic is also known as Outer Mongolia. (Inner Mongolia is the name given to the area of China that adjoins Mongolia.) The north of Mongolia is adjacent to the Siberian provinces of the USSR; its three other sides are bordered by Communist China. It is larger than Alaska.

Mongolia is changing from a nomadic culture to one of settled agriculture and small industries.

Population: 1.3 million. Area: 604,000 square miles. Capital· Ulan Bator.

Colors: deep purple, white, chartreuse

Laika, the Space Dog

The first flying zoo was demonstrated on September 9, 1783, in Paris, in the presence of King Louis XVI and his court. Étienne Montgolfier ascended in a beautiful golden balloon made of paper and cloth, taking with him a sheep, a rooster, and a duck. The only reported mishap was that at 6,000 feet, the sheep kicked the duck. Actually, the duck fared better than Laika, the first space dog.

Laika was orbited in Sputnik II on November 3, 1957, together with scientific instruments that transmitted to Earth reports on Laika's heartbeats, breathing, and other biological functions. Then, as described in "Events in Space" by Willy Ley, "After 100 hours in orbit, when the batteries in the satellite began to run down . . . the dog was painlessly put to death. . . ." Sputnik II burned up while reentering the Earth's atmosphere, carrying Laika, the first canine space hero, to a fiery end.

The Socialist Republic of Rumania, a Balkan state in southeastern Europe, is almost the size of Oregon. It is bounded by the USSR, the Black Sea, Bulgaria, Yugoslavia, and Hungary. Rumania has become a heavily industrialized country. Its main industries are metallurgy, machinery, oil, chemicals, timber, textiles, and food processing.

Population: 20 million. Area: 92,000 square miles. Capital: Bucharest.

Colors: pale turquoise, chocolate brown, white

Explorer I

January 31, 1958, was a big day in American space exploration. On that day the United States orbited its first satellite, Explorer I. It was only a modest satellite, some eighty inches long and six inches in diameter, weighing 18.13 pounds, but it marked the beginning of new dimensions for science, technology, and the industry of the nation. Although it was small, the information it brought back led to an important discovery: a belt of intense radiation above the Earth, starting at five hundred miles. It was filled with a higher radiation than had ever been detected above the atmosphere. The belt was named after its discoverer, James A. Van Allen of Johns Hopkins University, who analyzed the measurements made by the satellite's instruments.

Hong Kong is a British colony on the southern coast of China. Most of the colony's area is mainland territory, the rest is comprised of the 32-square-mile island of Hong Kong and dozens of tinier islands. Hong Kong is one of the busiest ports in the Far East.

Population: 3.9 million. Area: 398 square miles. Capital: Victoria.

Colors: carmine red, sky blue, black, white, silver

Luna III

Luna III was launched on October 4, 1959. This Russian satellite, a cylinder with a spherical bottom, weighed 613 pounds. It carried two cameras and other instruments, and it circled the earth in a long, cigar-shaped orbit. The far side of the orbit extended 50,000 miles past the moon's orbit, which made it possible for its cameras to take photos of the other side of the moon.

For over three hundred years, ever since Galileo invented the telescope, man had eagerly studied the visible side of the moon. Now, for the first time, a glimpse of the mysterious hidden side had become possible, as Luna III sent back televised pictures of it, as well as other information regarding the moon's environment.

Because the territory of the USSR is so vast, covering one sixth of the Earth's land surface, it contains both a varied climate and topography. The European portion is mainly an enormous low plain, broken by marshes and low hills; the Asiatic portion also consists of an immense plain, with deserts in the center and mountain ranges on its eastern and southern borders.

Colors: deep blue, white, mint green

Tiros Series

On April 1, 1960, the National Aeronautics and Space Administration (NASA) orbited Tiros I, first of a most successful satellite series. Tiros I was a meteorological satellite that delighted the world's weather forecasters by giving them their first look at the Earth from space. That first Tiros was followed by nine more, providing meteorologists with more than half a million photographs of cloud formations and adding a new dimension to weather forecasting.

The name Tiros is made up of the initials of *T*elevision and *I*nfra-*R*ed *O*bservation *S*atellite.

The Republic of Haiti occupies the western third of the island of Hispaniola in the Caribbean Sea. Though French is the official language, most Haitians speak Creole, a combination of African dialects, English, Spanish, old French, and Indian words.

Coffee is the chief product, along with sisal, cotton, sugar, bananas, tobacco, and rice. Major mineral exports are bauxite and copper. Haiti is a little larger than Maryland.

Population: 5.2 million. Area: 11,000 square miles. Capital: Port-au-Prince.

Colors: shamrock green, salmon, white

Cosmos Satellites

Cosmos I, the first of the Russian scientific series, was launched on March 16, 1962. It was designed to verify and improve space-vehicle construction, and to measure the effect of meteoric matter on space vehicles. Later Cosmos satellites were instrumented to study the Earth's magnetic field, the radiation belts, and solar and cosmic radiation.

Today hundreds of Cosmos satellites are orbiting the Earth. Some of them have electromagnetic surveillance and meteorological equipment aboard. They may also serve as navigational aids for submarines.

Czechoslovakia is a central European socialist republic about the size of New York State. The nation is highly industrialized, but agriculture remains important. There is a large glass and china industry. Other products include chemicals, aircraft, textiles, wood pulp. Chief crops are wheat, sugar beets, potatoes, rye, and hops.

Population: 15 million. Area: 49,000 square miles. Capital: Prague.

Colors: sandalwood, steel gray, charcoal, cream

Flying Zoo, USSR

The Soviet Union continued to experiment with animals in space. In March of 1961, two capsules were launched, each carrying both instruments and animals. Sputnik IX carried a dog, Chernushka. Sputnik X also had a dog, Zvezdochka, as well as guinea pigs, frogs, microbes, viruses, fungi, dry seeds, and onion shoots, all of which, after seventeen orbits, returned to Earth unharmed. Instrumented mannequins, to simulate astronauts, also were included in this flight.

The USSR is a federation of fifteen republics. Its area measures some 8.5 million square miles, over two times the size of the United States.

The Russian Soviet Socialist Republic, with its 130 million inhabitants, is the largest republic; the Estonian Soviet Socialist Republic is the smallest: population 1.3 million.

Colors: royal purple, black, sky blue, white

Space Puppy

When the Kennedys and the Khrushchevs met in Vienna in 1961, Mrs. Kennedy jokingly told Mr. Khrushchev that she had heard that one of the Russian space-traveler dogs had had puppies and that she would like to have one. Two weeks later the Russian Ambassador, with two aides, knocked at the White House door, bringing a puppy in a box.

Dog certificate No. V-087 read:

Breed: nonbreed. Name: Pushinka. Sex: female; Place of birth: the nursery of the USSR Academy of Science. Color: white, fluffy, with pale-yellow spots. Weight: six kilograms. Mother: Strelka, a space dog. Signed: Director of the Nursery Radkievitch.

The Kremlin in Moscow, ancient citadel of the Czars, forms the nerve center of the Soviet Union. Leningrad, formerly St. Petersburg, or Petrograd, is the second-largest city. Kiev, the capital of the Ukrainian SSR, is the industrial center of the south. The Crimea is a modern vacationland.

Colors: amethyst, sulfur yellow, white

Flying Zoo, USA

For both the Soviet Union and the United States a number of questions had to be answered before a man was sent into space. How will weightlessness affect his body and mind? How will man perform under stress? Can he adapt to the new environment of space? The answers to these questions were not only important for future cosmic flights but could have implications for health care on Earth. To find the answers man used cells, plants, cockroaches, dogs, and monkeys in space exploration.

On January 31, 1961, Mercury-Redstone 2 was launched from

Cape Canaveral with a passenger, a 37-pound chimpanzee named Ham. He returned safely after reaching an altitude of 157 miles in a suborbital flight preceding the first such manned flight. On November 29, a fellow chimpanzee named Enos survived two orbits in a Mercury-Atlas 5, in preparation for the three-orbit flight of John Glenn three months later.

Rumania is the world's sixth-largest corn producer. Also important are wheat, sugar beets, grapes, and other fruits. The country also has considerable mineral wealth, consisting of oil, natural gas, coal, salt, bauxite, manganese, lead, zinc, gold, and silver.

Colors: Prussian blue, white

Yuri Gagarin (1934–1968)

By 1961 a number of satellites were orbiting the Earth and reporting to scientists on various subjects, but no man had ever flown a space vehicle. Then, on April 12 of that year, the Soviets took a giant step.

On that day, the Soviet citizen Major A. Gagarin became the first man to fly an orbit around the Earth. A 125-foot rocket hurled his five-ton spaceship Vostok ("East") more than two hundred miles high at 17,000 miles an hour. After eighty-nine minutes in orbit, man's first flight in outer space had been successfully completed. Yuri Gagarin's flight was a milestone in the history of mankind. It was of great scientific and practical importance, for it showed us that man, under the right conditions, could endure the terrible stress of launching and lifting, feel quite normal during space flight, and safely overcome the dangers of returning to Earth.

Yuri Gagarin opened the way to outer space for the rest of the world, and he has received worldwide recognition for his exploit.

Later, Gagarin paid a courtesy visit to the United Nations in New York. He was an ever-smiling, compact young man who walked with a broad gait.

In 1968, Yuri Gagarin and a backup cosmonaut were killed in a plane crash.

The Mongolian People's Republic is one of the world's oldest countries. It reached the zenith of its power in the thirteenth century, when the Mongol leader Genghis Khan and his successors conquered all of China and extended their influence as far west as Hungary and Poland. In later centuries the empire dissolved, and Mongolia came under the suzerainty of China until the 1920's, when it became a Soviet satellite.

Colors: royal blue, aquamarine blue, black, white, crimson, violet, battleship gray

The Original Seven

By the end of 1958 the United States was ready to select its astronauts. The jobs were to be filled by candidates chosen from the existing pool of military test pilots. The candidates had to be under forty years of age, under five feet eleven inches tall, and in excellent physical condition; it was required that they had been graduated with a bachelor's degree or had an equivalent education, had been graduated from test pilot school, with fifteen hours minimum total flight time, and that they were qualified in jet aircraft. Although six men were needed, the final selections turned up seven names. They were M. Scott Carpenter, L. Gordon Cooper, John H. Glenn, Jr., Virgil I. Grissom, Walter M. Schirra, Jr., Alan B. Shepard, Jr., and Donald K. Slayton.

They became the "glamour boys" of the space age. Americans wanted to know all about them: What sports do they like? Which ones have been Boy Scouts? Who is their favorite movie star? How many children do they have? What do their wives think about the risk of space flight?

Before World War Two, Hungary was primarily agricultural, but industry has surpassed farming in value. About 70 percent of its foreign trade is with Eastern-bloc countries. Hungary is about the size of Indiana.

Colors: deep turquoise green, white, olive tan

Alan B. Shepard, Jr.

On May 5, 1961, a voice from a tiny American space capsule came crackling out of a loudspeaker: ". . . The fuel is go . . . oxygen is go. *Freedom Seven* is go." The voice belonged to Navy Lieutenant Commander Alan B. Shepard, Jr., America's first space hero, whose Mercury capsule, propelled by a Redstone rocket, rose 116.5 miles into the sky and reached a speed of 5,180 miles an hour. The fifteen-minute suborbital flight took

Shepard from Cape Canaveral to a point near Grand Bahama Island in the Atlantic. There, the spacecraft descended by parachute, and craft and pilot were picked up by helicopter and lifted to the waiting aircraft carrier.

Later Shepard said about the dangers of the venture: "There is no real point in worrying; all you need do is close out fear and concentrate on the job."

The figure 7 in the name *Freedom 7* represented the first seven astronauts.

In Paraguay the plains provide excellent pasturage and rich soil for agriculture; the mountain slopes are covered with luxuriant forests. Paraguay is one of the best-watered countries in the world, averaging sixty inches of rainfall a year. The most important crops are corn, cotton, rice, sugarcane, tobacco, and citrus fruits.

Colors: powder blue, crimson, white

Virgil I. Grissom (1927–1967)

Captain "Gus" Grissom of the Air Force made the second manned suborbital flight, the Mercury-Redstone 4, on July 21, 1961. His capsule, *Liberty Bell 7,* reached an altitude of 118 miles, to land 303 miles down-range sixteen minutes later.

One of the improvements made in the capsule since Shepard's flight was a larger window, and Grissom was fascinated by what he saw: the earth's curving horizon, which was ringed by bands of light and dark blue, and—as the capsule moved into the sun after passing through some clouds—a deep blue sky, which suddenly changed to black.

One of the innovations almost cost Grissom his life. A special explosive cord had been placed around the hatch, to enable the astronaut to leave the capsule quickly in an emergency. But as the spacecraft hit the ocean, with a "pretty good bump," the hatch door was blown off. Water came rushing in, and Grissom had to leave the capsule in such a hurry that he forgot to close a vent in his watertight suit. The water started to flow in and he began to sink. The helicopter hauled him up just in time.

The *Liberty Bell 7,* nearly filled with water by now, was too heavy to be lifted, and it sank to the bottom of the Florida Trench, in sixteen thousand feet of water.

The United States Mercury Project was a massive undertaking. It involved over two million people in its planning, preparation, and development, including hundreds of thousands of workers in the manufacture of the hardware and equipment needed. Project Mercury included twenty-five space flights and was the first major program of the National Aeronautics and Space Administration established in October of 1958.

Colors: Dresden blue, yellow, ocher, white

Gherman S. Titov

On August 6, 1961, a twenty-six-year-old Russian Air Force major, in a Vostok II capsule, circled the Earth seventeen times at an altitude alternating between 110 and 160 miles and returned safely to land in the Soviet Union. Americans splashed down in the sea; Soviet cosmonauts landed in the Soviet Union on dry land.

Titov later described the sensation of weightlessness: "For the life of me, I could not determine where I was. I felt suddenly as though I were turning a somersault, and then flying with my legs up. I was completely confused, unable to define where was Earth or the stars. . . . My sense of orientation vanished abruptly and completely."

Titov had become ill during his voyage, especially when he moved his head. This suggested that weightlessness interfered with the workings of the inner ear, which governs the sense of balance.

Vietnam, the easternmost nation of the Indo-China peninsula, was a French colony until 1954, when the Vietnamese defeated the French forces in a bitter battle at Dien Bien Phu. An international conference at Geneva then divided Vietnam along the seventeenth parallel. The industrial North came under Communist rule; the agricultural South, under a nationalist government. The two Vietnams have been at war since 1954, when North Vietnam, backed by Communist China, invaded the South. The United States sent military advisers and later troops in support of South Vietnam. In 1969, President Nixon began a "Vietnamization" program, under which United States troops were gradually withdrawn as South Vietnamese forces took over combat duty. Peace negotiations, which got under way in Paris in 1968, continued into the 1970's without any settlement being reached.

Colors: pale turquoise, white, coffee brown

President John F. Kennedy Looks into *Friendship 7*

February 20, 1962, was another big day in the history of American space exploration. On that day Marine Lieutenant Colonel John H. Glenn, Jr., riding in the capsule *Friendship 7,* orbited the Earth. The liftoff had been scheduled originally for February 14. Although the weather forecast promised fair skies, out at sea a storm was whipping up 10-foot waves. The flight was postponed 10 times. Glenn remained in good spirits, saying, "When we get the weather we will go."

The morning of the flight Glenn was awakened at 2:20 A.M. After a shave and shower he sat down to a breakfast of orange juice, a small filet mignon, and scrambled eggs, washing them down with Postum. Then he underwent a series of tests to record his respiration, blood pressure, heartbeat, and ability to perform coordinated movements. Glenn slid into *Friendship 7;* the white-dressed technicians closed the hatch. After a series of minor troubles were eliminated, the countdown and finally the magic word "Ignition" came. Orange flame erupted from the base of the Atlas rocket; the engines reached full thrust. On its pillar of fire and smoke and amid the thunder of engines, the first American was going into orbit, to circle the Earth three times.

Sharing the mood of elation felt by the whole nation, President Kennedy flew to Cape Canaveral to welcome John Glenn home.

Ras al Khaima is one of the seven sheikhdoms that make up the Trucial States. Located on the southern coast of the Persian Gulf, the total area of the Trucial States is 32,300 square miles—about the size of the state of Maine. The area has been a British protectorate for over a hundred years. Its people, mainly nomadic, are Arabs, Irani, Baluchi, and Indians. The total population of the seven sheikhdoms is 180,000.

Colors: steel blue, tan, cream

Rendezvous in Orbit

After the seventeen-orbit flight of Titov in Vostok II, the Russians flew four more Vostoks in two paired missions. On August 11, 1962, Major Andrian G. Nikolayev went aloft in Vostok III, and Lieutenant Colonel Pavel R. Popovich followed into orbit in Vostok IV on the next day. The two spacecraft, often in sight of each other, circled the Earth for several days. Both astronauts safely landed by parachutes in Kazakhstan.

Two large mountain ranges, the Carpathians and the Sudetes, run along the southern border of Poland. The Masurian lake region in the northeast is dotted with some three thousand lakes.

Education is free from nursery through university, and compulsory from seven to sixteen years of age.

Colors: wisteria blue, chartreuse green, white, black

First Woman in Space

The next Soviet double mission followed in June, 1963. Lieutenant Colonel Valery F. Bykovsky flew into orbit in Vostok V; two days later the first woman astronaut, Valentina V. Tereshkova, followed him aboard Vostok VI. The two spacecraft at one point came within three miles of each other. The two cosmonauts landed safely two and a half hours and fifty miles apart.

Valentina Tereshkova later married cosmonaut Andrian G. Nikolayev, and she is the mother of a girl. When the American astronaut Neil Armstrong, the first man on the moon, visited Moscow in 1970, it was cosmonaut Nikolayeva Tereshkova who pinned a souvenir badge on his lapel.

The "DDR" shown on the stamp stands for "Deutsche Demokratische Republic" (German Democratic Republic).

The German Democratic Republic was established as a Soviet-controlled state after World War Two. Although much poorer in natural resources than West Germany, the DDR ranks today among the ten greatest industrial countries in the world. Major industries include shipyards and modern chemical and electronic plants. It is especially famous for its optical instruments.

Colors: bottle green, mint green, ultramarine, white

More Americans in Space

After John Glenn, more Americans flew into space. On May 24, 1962, Navy Lieutenant Commander M. Scott Carpenter orbited the Earth three times and landed 250 miles off target. He was followed, on October 3, 1962, by Navy Commander Walter M. Schirra, who completed six orbits. The Mercury Program was closed out by Air Force Major L. Gordon Cooper's twenty-two-orbit flight on May 15, 1963. Cooper made a pinpoint landing, coming down 4.5 miles from the recovery ship.

The Mali Republic, one-time French overseas territory in West Africa, is larger than Texas but smaller than Alaska. From the eleventh to the fifteenth century the area was part of the great Malinké empire. Timbuktu was a center of trade and of Islamic learning.

As a colony of France it was known as the French Sudan. After independence, Sudan and Senegal formed the Federation of Mali in 1959. After Senegal withdrew from the federation in 1960 Sudan proclaimed itself the Republic of Mali.

Population: 5.1 million. Area: 465,000 square miles. Capital: Bamako.

Colors: orange, dark turquoise, white, light and dark gray, buff

Three Men in a Spacecraft

The Russians once more led the way when, on October 12, 1964, they orbited the first of their Voskhod ("Sunrise") series with three men aboard: Colonel Vladimir M. Komarov, the flight commander, and Boris G. Yegorov and Konstantin P. Feoktistov, two civilian scientists who conducted experiments. During part of the flight the cosmonauts were dressed in "shirt sleeves," rather than the bulky space suits. After sixteen orbits the cosmonauts landed with their craft, instead of parachuting during the descent, as previous cosmonauts had done.

The Republic of the Niger, in the heart of West Africa, is a vast plateau, almost twice the size of Texas. Niger, a former French colony, is an agricultural and pastoral land. Peanuts are the chief product. Camels, donkeys, sheep, and goats are also important.

Population: 3.8 million. Area: 489,000 square miles. Capital Niamey.

Colors: sapphire blue, dark grayish turquoise, carmine, white

Scientific Satellites

Both the Soviet Union and the United States continued to orbit scientific satellites while men were orbiting the earth.

In 1958, the United States launched the Explorer and Vanguard series. They supplied valuable information about the Earth and its space environment, and confirmed the existence of the Van Allen radiation belts and of the "pear shape" of the Earth. The Air Force's Discoverer program, begun in 1959, brought back important solar and cosmic radiation data and helped develop the technology needed for recovery of packages sent to Earth from orbiting satellites.

In July of 1962, the United States launched Telstar, the first communications satellite. Communications satellites were placed in orbit to pick up telephone, telegraph, and television signals from one point on Earth and then send them down to another point. Because these satellites are so high above the Earth and their signal range is from horizon to horizon, they can relay television programs between distant points. Telstar, for example, transmitted a television show live from France to the United States. It could amplify ten billion times a radio signal received from the ground, and return it on another frequency.

The Soviet Union launched its first Molniya ("Lightning")

communications satellite in April, 1965; through the spring of 1969, a total of eleven were orbited.

That country started its Proton series in 1965 with a huge 27,000-pound satellite, followed by three more in the next five years. Their announced purpose was to study high-intensity radiation particles, but their huge size led to speculation that they were prototypes of manned space stations.

The Federal Republic of Cameroon on the west-central coast of Africa is larger than California. There are about two hundred tribal groups, whose main divisions are Bantu, Sudanese, and Arab-Foulbé. About a third of the population is Christian, half that number Moslem, and the rest, animist. While French and English are the official languages, twenty-four other African languages are spoken.

Cameroon exports coffee, cocoa beans, aluminum, wood, cotton, and rubber.

Population: 5.8 million. Area: 183,000 square miles. Capital: Yaoundé.

Colors: forest green, white, lobster, sapphire blue

Syncom Satellites

The word "Syncom" is composed of *syn*chronous and *com*munication, and the purpose of a Syncom is to provide a link between telephone, radio, and television communications. Since its orbit is synchronized with the rotation of the Earth, a Syncom satellite remains "stationary" over a precise geographical point. Syncom I was launched by the United States on February 13, 1963. It went into orbit, but its radio equipment failed to work. Syncom II and another Satellite, Relay I, were used to link Rio de Janeiro, New Jersey, and Nigeria, in a three-continent conversation.

Syncom III was placed into orbit on August 19, 1964. It achieved a synchronous orbit 22,300 miles above the equator over the Gilbert Islands.

The Republic of Senegal, a former French overseas territory, is on the Atlantic coast of western Africa. It is about as large as

South Dakota. Peanuts are the mainstay of the economy of the country. Dakar, the capital, is an important seaport.

Population: 3.9 million. Area: 76,000 square miles. Capital: Dakar.

Colors: sky blue, white, Indian red, clover green

France's Satellites

The United States and Russia are not the only nations with space programs. In April, 1962, an American rocket orbited Ariel 1, a 132-pound satellite carrying British-built scientific instruments. The Italians built the San Marco 1 satellite in 1964, and the French built FR-1 in 1965; both used American carrier vehicles. The French achieved their own space program with the successful orbiting of satellites launched from the Sahara and French Guiana. Because these locations are closer to the equator, where the Earth's rotational speed is greater, they are more suitable for placing satellites into equatorial orbits.

The Islamic Republic of Mauritania is a former French colony in West Africa. It is about four fifths the size of Alaska. Products include dates, grain, tobacco, and fish. There are large herds of cattle, camels, sheep, and goats. Mauritania has some of West Africa's largest deposits of iron ore and copper.

Population: 1.2 million. Area: 398,000 square miles. Capital: Nouakchott.

Colors: marine blue, Indian red, white, royal purple, reddish brown

Canada's Alouette

On September 28, 1962, a Canadian atmospheric research satellite was launched. Alouette 1 was the first satellite completely designed and built by a nation other than the United States or the Soviet Union. It is one of dozens of *topside sounders,* which explore the upper atmosphere.

International cooperation in space exploration is being conducted by the European Space Research Organization (ESRO), which, with NASA-provided carriers, launches satellites to study solar and cosmic radiations and ionospheric and aural phenomena in the polar regions.

Canada is a federation and a constitutional monarchy. The head of the state is Queen Elizabeth II. The queen is represented in Canada by a resident governor-general appointed by Her Majesty on the advice of the Canadian federal cabinet.

Statistics indicate that Canada has the world's second- or third-highest standard of living.

Population: 22 million. Area: 3,852,000 square miles. Capital: Ottawa.

Colors: midnight blue, white

Lunar Exploration

The Soviets started lunar exploration by firing their first three Luna spacecraft in 1959. Luna I was launched on January 2, 1959. It passed within 2,187 miles of the moon's surface and then flew on to orbit the sun. It broadcast data when it was as far as 373,125 miles from the Earth. The United States did essentially the same thing two months later with Pioneer 4.

On September 12, the Russians launched Luna II, which carried 858 pounds of instruments. It became the first man-made object to hit the moon, crashing on the surface, but sending back signals before impacting.

Luna III, launched on October 4, circled behind the moon and sent back pictures that gave man his first view of the far side of the moon.

The constitution of the Democratic Republic of Vietnam is based on Communist principles and calls for the reunification of all Vietnam. It provides for a president elected by parliament and a prime minister appointed by the president. Since the death of President Ho Chi Minh in 1969, a collective group has been leading the nation.

Colors: fiery orange, white, black

The Lunas and the Rangers

A number of Soviet Lunas crashed on the moon, but finally, in January of 1966, Luna IX soft-landed, sending back a series of closeup pictures demonstrating that the moon's pockmarked face was hard enough to support the weight of a manned spacecraft.

The Soviet Union was ahead in the lunar race until Ranger 7, launched July 28, 1964, sent back 4,316 pictures of the moon, including some that showed craters that the best telescopes had been unable to see. Ranger 8, in February, 1965, sent back 7,137 pictures. Ranger 9, launched March 21, 1965, gave Americans the thrill of seeing live television pictures from the moon on their home sets.

Education in Rumania is compulsory for ten years, and all education is free. The Rumanian language has a Latin base, with traces of French, Greek, Slav, and Turkish influences.

Since 1965, Rumania has called itself a Socialist, rather than a People's, Republic. It maintains friendly relations with the USSR as well as with the USA.

Colors: deep blue, white, sepia, orange, black, amethyst

Moon Rush

By 1966 the unmanned attempts to explore the moon were in full swing. In June the American spacecraft Surveyor 1, after a successful soft landing, televised photographs to Earth. The Russian Luna XIII, which soft-landed on the moon, on December 21, 1966, sent back close-up surface photos of the moon. Explorer 35, of the United States, studied solar winds, small particles, and the magnetic field in lunar space. The Russian Luna XIV, which landed on the moon on December 24, 1966, studied the relationship between the masses of Earth and moon; in January, 1968, Surveyor 7, the last of the American series, returned 21,000 photos of the moon, the Earth, and Jupiter.

Fujeira is one of the seven sheikhdoms of the Trucial States. Each of these tiny states had one major city, which has the same name as the state and is its capital. The seven sheikhs form the Trucial States Council, the ruling body for the area. The main industry—aside from the oil in one state, Abu Dhabi—is fishing. Dates and tobacco are the chief crops, and pearls are also exported.

The term "trucial" refers to truces made by the seven sheikhs with Great Britain in the nineteenth century. The area was formerly known as the Pirate Coast.

Population: 180,000. Area: 32,000 square miles.

Colors: magenta, blood red, white, black, sky blue, yellow

Unmanned Flight Around the Moon

The Russians launched Zond 5 on September 15, 1968, to orbit around the moon. It was guided automatically back to Earth and splashed in the Indian Ocean. Zond 5 carried a collection of worms, flies, seeds, plants, and turtles, which were later examined to determine the effect on living organisms of radiation beyond the Earth's magnetic field. Zond 5 was followed on November 10 by Zond 6, which also carried biological specimens and took pictures of the moon.

Zond 7, launched August 8, 1969, used a heavy booster rocket, similar to those used to launch manned space flights. It went around the moon without orbiting and then returned to Earth.

The Communist party is the only legal party in the USSR. Its highest organ is the Party Congress of about fifteen hundred elected representatives, which normally meets once every four years.

The USSR and the US signed an agreement in 1963 establishing an instantaneous Teletype-and-radio communication between the Kremlin and the White House. It is known as the "hot line" and designed for emergency communications and as a safeguard against accidental war.

Colors: black, white, mocha, blue-gray

Space Walk

Once more the Russians scored with a "first."

On March 18, 1965, Lieutenant Colonel Aleksei A. Leonov left Voskhod II through an airlock and spent ten minutes in space while the other crew member, Colonel Pavel I. Belyayev, watched him.

The airlock was mounted to permit access from the cabin through a hatch, which was activated by an electric drive mechanism. Once in the airlock, the cosmonaut could open the outer hatch either manually or electrically. Cameras automatically recorded his activities, and more cameras were mounted on the exterior, as were air tanks for pressurizing the airlock after he returned from space. The airlock was jettisoned after he entered the cabin.

Two and a half months later the United States matched the Russian space walk, which became known as an "extravehicular activity," or EVA.

Mali is mainly agricultural and pastoral. Millet, rice, and

peanuts are the chief crops. Cotton, rubber, and river fishing are important. Livestock raising is a major factor of the economy.

Colors: *peacock blue, dark blue, orchid pink, pale lavender, light and dark gray, ocher, bright red, white, black*

Gemini Series

The American manned space flight program started with Project Mercury, whose object was to place one man in orbit around the Earth. Six astronauts participated in the Mercury program: Alan Shepard, Virgil Grissom, John Glenn, Scott Carpenter, Walter Schirra, and Gordon Cooper. Donald Slayton, one of the original seven, was unfortunately grounded for cardiac reasons (irregular rhythm), but he remained with NASA, where he became director of Flight Crew Operations.

Project Mercury was considered a prelude to the Apollo program, which ultimately was to take man to the moon. Mercury had proved a lot about space flight, but not enough to serve as a basis for the Apollo program.

The Gemini program grew out of this need. It was the transitional stage between the Mercury and Apollo programs.

It provided the long-duration experience, the experience needed for orbital rendezvous and for extravehicular activities, or space walks, and it improved landing techniques and ground-crew skills.

Niger became fully independent from France on August 3, 1960, and joined the United Nations on September 20 of that year. It elected to remain outside the framework of the French community, but signed a bilateral agreement retaining close ties with France. The republic has a president and a National Assembly, elected for five-year terms.

Colors: sapphire blue, dark violet, white

Extravehicular Activities (EVA's)

Gemini is the third sign of the zodiac meaning "twins." In American space language, it meant that the spacecraft carried two astronauts.

The first manned Gemini flight took place on March 23, 1965, five days after Voskhod II was launched. Virgil Grissom and Navy Commander John W. Young made three orbits in the Gemini spacecraft *Molly Brown,* using the on-board rockets to change the size and shape of their orbits.

Majors James A. McDivitt and Edward H. White II rode Gemini IV for sixty-two orbits. White left the cabin to spend twenty-one minutes floating in space, maneuvering with a gas gun in hand. It was the first time an astronaut could exercise control over his movements in space with a hand-held device.

The United States captured the space-endurance record in August of that year, when Gordon Cooper and Charles Conrad, Jr., in Gemini V, orbited for seven days, 22 hours, and 59 minutes.

Nicaragua is the largest of the Central American states. Although it is essentially an agricultural country producing bananas, cotton, fruits, coffee, sugar, cocoa, rice, and tobacco, industrialization is growing.

Population: 3 million. Area: 50,000 square miles. Capital: Managua.

Colors: azure blue, grass green, yellow, red, white, black

Gemini IV

When the flight of Gemini IV was first programmed, there was still no EVA suit. One was adapted from the flight suit designed for Project Gemini. The suit, constructed of twelve layers of rubber, felt, nylon, etc., created a microenvironment in which a man could live in space. It pressurized his skin so that the internal pressures of his body would not rupture his blood vessels, it retained his body heat so that he could not be frozen stiff, and it enabled him to breathe and get rid of some of his excess body moisture.

Project Gemini was the intermediate step in the American space effort to put a man on the moon. Much of the preliminary work and much of the responsibility for these flights were centered in the Manned Spacecraft Center (MSC) in Houston, Texas. The MSC was built at a cost of $1.5 billion and comprises twenty very modern buildings. Part of MSC is an acre of ground that has been made into a "lunarscape" for testing models of equipment to be used on the moon.

Colors: navy blue, sky blue, white, turquoise, cherry red

20
DDR
MONOMATERIE ZUR ERDE ZURÜCK
MONDSATELLIT LUNA 10
SOJUS 4 U. 5
SOJUS 9
NACHRICHTENSATELLITEN MOLNIJA 1
WOSTOK-RAKETE
BOCTOK
CCCP
12.4.1961
I. A. GAGARIN MIT WOSTOK 1
18.3.1965 AUSSTIEG A. LEONOW AUS
16.5.1969 VENUSSONDE

Docking

On March 16, 1966, civilian pilot Neil A. Armstrong and Air Force Captain David R. Scott coupled Gemini VIII and an Agena target vehicle in the fifth orbit, thus accomplishing the first docking operation in space.

Gemini IX, launched June 3 with Air Force Lieutenant Colonel Thomas P. Stafford and Navy Lieutenant Eugene A. Cernan as pilot, underscored that an astronaut's lot in space is not a happy one. Cernan's activities were so demanding that he lost about ten pounds during the flight and two pounds of perspiration were found in his space suit after recovery. His respiration rate doubled in space.

During the flight Cernan took a space walk that lasted more than two hours.

The twelve Gemini flights built up the total of EVA time by American astronauts to about thirteen hours.

While the space walks were the most dramatic features of the Gemini programs, each flight accomplished many different missions, including the study of the effects of prolonged weightlessness and the taking of color photographs of the Earth's land and water to help geologists, oceanographers, and meteorologists. The beautiful stamps on this page illustrate various docking maneuvers.

Furthermore, there was a series of medical experiments and measurements to determine whether the heart and bones are weakened and mental ability is impaired by long space flights.

After World War Two, the western portion of Germany was occupied by the United States, Britain, and France; the eastern part, by the Soviet Union. In 1949 the country was officially divided into the Federal Republic of Germany (West Germany) and the German Democratic Republic (East Germany). Each now is recognized as a sovereign state. In East Germany the Volkskammer (People's Chamber) votes on the laws of the land.

Colors: sky blue, midnight blue, white, black, light gray, cherry red, yellow, ocher, silver, turquoise

Tragedy on the Launch Complex

After an eminently successful American space program that had seen six manned flights in the Mercury program and ten manned missions in the Gemini program, carrying twenty-four astronauts into space (some of them more than once), tragedy struck at Cape Kennedy. On January 27, 1967, astronauts Virgil I. Grissom, Edward H. White, and Navy Lieutenant Roger B. Chaffee died in a flash fire that destroyed an AS-204 spacecraft.

The three Americans were to have been the first astronauts in the Apollo manned-flight program. They would have been boosted into Earth's orbit by a Saturn 1B carrier vehicle on February 21, for a two weeks' trip.

While the astronauts sat in the vehicle, minor troubles had been corrected. Then a voice came from the spacecraft, crying "Hey," or "Fire." Two seconds later a voice was heard: "I've . . ." or "We've . . . got a fire in the cockpit." For another 6.8 seconds there was no further word from the cockpit. Then another garbled transmission came through: "They're fighting a bad fire. . . . Let's get out. . . . Open 'er up. . . . We are burning up. . . . I'm getting out!" The transmission ended with a cry of pain.

The Republic of Guinea, in western Africa, has a variety of climates, from the humid coastal tropics to cooler uplands. Wildlife includes elephants, buffaloes, hippopotamuses, antelopes, lions, leopards, chimpanzees.

Population: 3.9 million. Area: 95,000 square miles. Capital: Conakry.

Colors: scarlet, cream, light gray, salmon, black, gold, flesh color

Death of a Cosmonaut

In the spring of 1967, some two years after the Voskhod II mission, the Russians resumed manned space flights. Their Soyuz I was launched on April 23, in what probably was planned as a rendezvous flight with other cosmonauts. However, the spacecraft developed trouble, and after twenty-six hours in space Cosmonaut Vladimir M. Komarov was ordered to terminate the flight. During the reentry period the parachute failed to function and the capsule crashed, killing Komarov. This was the first known death of a space flier during the course of an actual mission. A monument was erected on the steppe near Orenburg, where cosmonaut Komarov met his death.

Manama is the capital of Bahrain, a British protectorate in the west central Persian Gulf between Qatar and Saudi Arabia. It is a sheikhdom made up of a number of desert islands inhabited, for the most part, by Moslem Arabs. The total land area is 231 square miles. Oil is its main industry, and there is a refinery on one of the islands, Sitra, which processes Arabian as well as local oil. The population of Bahrain is 200,000.

Colors: aquamarine, steel blue, sky blue, white, black, gray, brown, red, yellow, flesh color

Apollo 7

Apollo 7, the first manned test of a moonship, was lifted into orbit around the Earth on October 11, 1968, with Walter Schirra, Air Force Captain Donn F. Eisele, and civilian scientist-pilot R. Walter Cunningham.

The many experiments assigned to Apollo 7 included comprehensive terrain photography of the Earth's land and sea areas for the benefit of geoscientists and oceanographers. Global weather patterns were also photographed. In addition, a number of medical tests were conducted, including studies of blood, bone demineralization, and the effects of prolonged periods of weightlessness on the heart.

Apollo 7 accomplished another "first." The television coverage of the activities of the astronauts aboard the craft was live, and millions watched it.

All three astronauts developed colds during the flight. They were given medical advice via radio from the ground station.

Qatar, a sheikhdom peninsula on the Persian Gulf on Saudi Arabia's northeastern coast, had been under a British protectorate since 1916. On September 1, 1971, Qatar became an independent state.

The coast offers the most productive pearl fishing in the world. But countless shoals and reefs render the navigation dangerous. Storms, too, often arise suddenly, and shipwrecks are a frequent occurrence.

Population: 80,000. Area: 4,000 square miles. Capital: Doha.

Colors: salmon, muted blue, black, white, vermilion

Meeting with Santa Claus

Apollo 8, with Air Force Colonel Frank Borman, Air Force Captain William A. Anders, and Navy Captain James A. Lovell, Jr., aboard, was scheduled to be launched Saturday, December 21, 1968, four days before Christmas.

It was to fall into an orbit around the moon. After ten revolutions of two hours each, just sixty-nine miles above the moon's surface, the service propulsion system engine would be fired to accelerate the spacecraft out of lunar orbit and return it to the flight path back to Earth.

Almost immediately upon arrival in lunar orbit, the astronauts commenced photographing the spectacular scene beneath them. We learned that the surface of the moon is a grayish white, and that the craters look as though they had been made by "meteorites or projectiles of some sort." Live television broadcasts were conducted directly from the astronauts in lunar orbit. They became the first human beings to observe the far side of the moon, which seemed more rugged than the near side.

Anders' seven-year-old son Gregory wanted to know when Daddy would meet Santa Claus. Back came the answer: "We saw him earlier today and he was heading your way."

When the ship emerged from the far side of the moon, Mission Control erupted in cheering. The Apollo engine had fired as programmed, and the first men to go around the moon were safely on their way home. And Borman was saying, "Roger. Please be informed there *is* a Santa Claus!" (Had the engine failed to fire at this crucial moment, the men would have been doomed, their spacecraft locked in lunar orbit forever.)

Paraguay gained its independence from Spain in 1811. Its present constitution took effect in 1967 and provides for a Senate of thirty members and a Chamber of Deputies of sixty. Like all Latin American countries, Paraguay is a member of the United Nations.

Colors: mulberry, cinnamon, light tan, white

Genesis

On Christmas Day, Borman announced: "The crew of Apollo 8 has a message that we would like to send you," and Anders began reading from the first book of the Bible, Genesis.

"In the beginning, God created the heaven and the earth. And the earth was without form, and void; and darkness was upon the face of the deep. And the spirit of God moved upon the face of the waters. And God said, Let there be light; and there was light. And God saw the light, that it was good; and God divided the light from the darkness."

Now Lovell took over: "And God called the light Day, and the darkness He called Night. And the evening and the morning were the first day. And God said, Let there be firmament in the midst of the waters and let it divide the waters from the waters. And God made the firmament, and divided the waters which were under the firmament from the waters which were above the firmament; and it was so. And God called the firmament Heaven. And the evening and the morning were the second day."

Then Borman spoke: "And God said let the waters under the Heaven be gathered together in one place and let the dry land appear; and it was so. And God called the dry land Earth; and the gathering together of the waters He called the Seas; and God saw that it was good."

And from the crew of Apollo 8 came, "Good night, good luck, a merry Christmas, and God bless all of you—all of you on the good Earth."

Colors: black, white, sepia, pale blue

Which Way to the Moon?

While the Soviet Union experimented with automatic scouts, the target of the American space effort was to land men on the moon. Several modes of travel were suggested. One was a direct ascent to the moon by means of a missile shot from Earth to moon. Another suggestion was the Earth orbit rendezvous (EOR). It proposed a rendezvous of two vehicles, launched at different times, in the Earth's orbit for the purpose of refueling the moonbound rocket. A third possibility was a rendezvous on the lunar surface. A fuel-supply vessel controlled by radio could be landed on the moon's surface, from which the manned lunar vehicle could take on fuel for the journey home—provided the astronauts could find the supply tank.

And finally there was the lunar-orbit rendezvous (LOR). It did not provide for the mother ship to land on the moon at all. Instead, two astronauts would descend to the lunar surface in a space "rowboat" while the mother ship with the third astronaut aboard remained in lunar orbit. The "rowboat" was known as the lunar excursion module, or LEM, later shortened to LM. However, because of its spindly landing legs, which resembled an insect's, it became better known as the *Bug,* or *Spider.* The command module (CM) came to be called *Gumdrop* for its cone shape and the blue plastic cellophane wrapping in which it arrived from the plant, which gave it a candy-like appearance. LOR, this last plan, was the mode of travel finally adopted for landing on the moon.

The pearlfishing season in Qatar begins in the spring and lasts all through the summer. The Arab pearlfisher can remain under water for two minutes at a time. The oysters are opened on shore and the pearls carefully sorted and counted. Some of the pearls are of great size and value.

Colors: chocolate brown, salmon, light sea-green, salmon-orange, primrose yellow, white

Apollo 9

Apollo 9 was launched on March 3, 1969, with James A. McDivitt in command, and with David R. Scott and Russell L. Schweickart, a civilian aeronautical engineer who was called Rusty—not for "Russell" but for his red hair and freckles.

Apollo 9 was a dress rehearsal for the landing on the moon. It carried aboard a lunar module ready for the astronauts to ride in.

McDivitt and Schweickart sat in *Spider,* and after fifty-nine revolutions of the Earth, Scott released *Spider,* which was then free of *Gumdrop.* For nearly two hours the two craft circled the Earth, reaching a maximum distance from each other of fifty miles. Besides the successful separation, rendezvous, and docking of *Spider* and *Gumdrop,* the mission accomplished a walk in space, Earth resources photography, space-suit and backpack checking guidance, and a computer and communication system between *Spider* and *Gumdrop.* It was a dramatic demonstration that man could control spacecraft in orbit. Apollo 9 was in orbit a total of 241 hours.

Of the rivers of Rumania, by far the most important is the Danube, which is navigable for large vessels throughout that country.

The climate of Rumania is one of extremes. In winter the thermometer may sink to minus fifteen degrees Fahrenheit, while in summer the temperature may reach ninety to nine-five degrees.

Rumania produces valuable trees—oaks, firs, and beeches. The country's warm summers and cold winters have a beneficial effect on the quality of the timber, for the wood is hard and durable.

Colors: azure blue, turquoise, black, white, silver

The Most Expensive Suit

In case you want to buy a space suit, the cost of manufacture is $2 million. Of course you get a few extras: a backpack containing oxygen, and air conditioning and a communication outfit. During the flight of Apollo 9, the space suit and backpack were checked out in space. Schweickart's oxygen tank contained a pound of oxygen. A ventilating circuit forced the oxygen through the suit and the backpack to remove poisonous carbon dioxide and contaminants, and to cool and remove perspiration. The moisture was then removed by a water separator. Water-transport systems kept the astronaut cool in spite of his own body-heat buildup and any heat leaking into the suit from the outside. The expendable water was forced outside the suit and vaporized.

A communications system included two radios, medical-data transmitters, and a warning system to the ground should something go wrong.

Like princesses in fairy tales, the astronauts have golden shoes—gold-coated, that is, to reflect the dazzling light of the sun and protect their feet.

Iran, formerly called Persia, is a constitutional monarchy in southwest Asia, slightly larger than Alaska. For neighbors it has the USSR, Afghanistan, Pakistan, Iraq, and Turkey. Iran is the world's fourth-largest oil producer. Agriculture is a prime industry, producing wheat, barley, corn, rice, yams, wool, tobacco, and raw silk. Persian carpets are world famous. Sturgeon fishing in the Caspian Sea is important as a source of caviar.

Population: 28 million. Area: 636,000 square miles. Capital: Teheran.

Colors: Persian blue, burnt umber, medium gray, ocher, white

Apollo 10

With the launching of Apollo 10 on May 18, 1969, new words were added to the list of the already rich space vocabulary. The command module was baptized *Charlie Brown,* after the beloved comic-strip character, and the lunar module *Snoopy,* after Charlie Brown's dog, no doubt in reference to the fact that the lunar module would soon be "snooping" on the surface of the moon.

The voyagers in *Charlie Brown* were Thomas P. Stafford, commander, big, baldish, with a slow Oklahoma drawl, a veteran of Gemini III, VI, and IX. His fellow pilot in *Snoopy* was Eugene A. Cernan. The man who was to remain in *Charlie Brown* while Stafford and Cernan sat on the back of *Snoopy* was John W. Young.

The mission of Apollo 10 included:

a) Flying the lunar module within nine miles of the surface of the moon, the closest man had ever gone to another celestial body.

b) Accomplishing a rendezvous and linkup between the lunar module and the command module about 70 miles from the moon and almost 250,000 miles from Earth.
c) Determining what effect the moon's weak gravitational field had on the accuracy of flight maneuvers.
d) Photographing Apollo 11's intended landing site on the Sea of Tranquility.
e) Transmitting the first color television pictures from deep space.

About 80 percent of the Rumanian people are members of the Rumanian Orthodox Church; most of the rest are almost equally divided between the Greek Orthodox Church and the Roman Catholic Church. The Rumanian Orthodox clergy are paid by the state, and other clergy also receive subsidies; but church and state can be called separated.

Colors: marine blue, light blue, salmon, white, black, silver

Snoopy and *Charlie Brown* Hugging

After checking all instruments to make sure they had survived intact the 250,000-mile journey, Cernan and Stafford crawled through the connecting tunnel into *Snoopy,* leaving Young in *Charlie Brown.* The two astronauts separated the lunar module, flew in formation for a while, and then, firing the lunar module's descent rocket, streaked down through the lunar sky at a speed of 3,700 miles an hour. *Snoopy* was within 49,000 feet of the moon's surface, about the altitude at which a jet flies over the Earth.

Stafford described the Sea of Tranquility, a prospective landing site, as "very smooth, like wet clay, like a dry riverbed . . ." The moon's lack of atmosphere meant there were no clouds or haze.

The astronauts fired the LM's ascent engine to boost it back toward their rendezvous with the command module. As the two craft were joined, one astronaut shouted, "*Snoopy* and *Charlie Brown* are hugging each other."

After Stafford and Cernan crawled back into *Charlie Brown,* they closed the hatches and jettisoned the lunar module that had brought them back.

The total journey took eight days, including two and a half days in orbit around the moon.

Apollo 10 was the first flight that proved the feasibility of lunar-orbit rendezvous, and the first to take man to the surface of the moon. Apollo 10 also served as a test of the LM's descent propulsion system not only in the lunar landing configuration but in the lunar environment. Apollo 10 was the lunar dress rehearsal for Apollo 11, which landed man on the moon.

Guinea is still an agricultural country, but the importance of minerals to its economy is growing. Bauxite, iron, diamonds, both gem and industrial, are produced.

Colors: scarlet, cream, light gray, black, gold, flesh color

Apollo 11

The three men selected for the journey to the moon in Apollo 11 were Neil A. Armstrong, Edwin Aldrin, an Air Force major, and Michael Collins, an Air Force captain.

The launch day was July 16, 1969. Nearly one million tourists had streamed into the communities around Cape Kennedy. They wanted to tell their children that they had seen the first men go to the moon. Pope Paul VI had arranged for a special color television set to be installed in his summer residence so he could watch the flight.

For the two months before the flight, each of the astronauts had spent more than four hundred hours in command-module and lunar-module simulators at Cape Kennedy and at Houston.

It was thus possible to practice every step of the mission—the launching, Earth-orbit operations, the firing of the rocket that enabled them to leave the Earth orbit and point the spacecraft toward the moon, the insertion into lunar orbit, the landing on the moon, the liftoff from the moon, and the return to Earth.

All three astronauts also had experienced a lunar vacuum in the large vacuum chamber at Houston, Texas.

The big moment arrived at 9:32 A.M. Eastern Daylight Time on a Wednesday. Orange flame and a cloud of black smoke lifted the 3,817-ton vehicle of the astronauts into their eight-day, 5,000-mile journey. Communications satellites relayed the launching for international television for the benefit of an estimated half billion people in the world.

Large grants from both Communist and non-Communist countries accelerated Guinea's agricultural output. Chief exports are aluminum, iron ore, and agricultural produce such as corn, palm nuts, coffee, and honey. Rice is the staple food of the population.

Colors: chrome yellow, salmon-pink, muted orange, white, black, gold, Indian red, tan, marine blue, light green

Columbia and *Eagle*

During the flight, the command module and the lunar module, this time called *Columbia* and *Eagle,* separated. *Columbia* turned around and docked with *Eagle,* and they continued the journey in this position. On July 19, the main rocket was fired to slow down the aircraft, so that it could enter the moon's gravity and go into orbit.

The next day, Armstrong and Aldrin, riding the *Eagle,* descended toward the Sea of Tranquility, the landing site. A myriad of craters, hills and ridges, cracks and rilles awaited them in the brown-gray moon landscape.

The *Eagle* closed in on a floor littered with a large number of boulders and rocks, a dangerous place for a landing. Armstrong took control of the lunar module and manually guided it while he searched for a clear spot for ninety seconds. Finally he found one just outside a crater. Slowly the *Eagle*'s legs touched the surface, and Armstrong's voice told Houston and the millions who were anxiously waiting to hear the news: "The *Eagle* has landed." Ground control replied, "You've got a bunch a guys about to turn blue. We're breathing again. Thanks a lot."

On July 13, the Soviet Union had launched an unmanned

moon-orbiting spaceship, Luna XV. The Russians assured the United States by cable and telephone that their spaceship would not interfere with Apollo 11. This was the first time that the Soviet Union had communicated directly with the Americans about one of their missions while it was in flight.

The Republic of Korea, "Land of the Morning Calm," is about the size of Indiana. It occupies a mountainous peninsula in northeast Asia separating the Yellow Sea from the Sea of Japan. The Republic was formed by South Koreans in May of 1948.

South Korea is chiefly an agricultural country. The main crops are rice, barley, wheat, and beans. The division of Korea in 1945, into north and south left the south with only light industry.

Population: 34 million. Area: 38,000 square miles. Capital: Seoul.

Colors: navy blue, white, vermilion, sky blue

". . . One Giant Leap for Mankind"

At 10:56 P.M. Eastern Daylight Time on July 20, 1969, Neil Armstrong made the first human footprint on the moon. As his left foot touched its surface, he uttered the immortal words, "That's one small step for a man, one giant leap for mankind." For the scientists this meant an unprecedented opportunity to obtain possible clues to the origin and nature of both the moon and the Earth.

After eighteen minutes, Aldrin joined Armstrong on the moon. The astronauts set up a television camera, and hundreds of millions of people on Earth witnessed their activities on the moon. Ironically, the man who was the nearest to the astronauts could not see what was happening; since *Columbia* was not equipped with a television receiver, Michael Collins had to be content with listening to his fellow astronauts' words.

In 1958, Guinea seceded from France and proclaimed itself an independent republic. The nation's first constitution was adopted the same year. Guinea is a member of the United Nations. French is its official language.

Guinea's constitution provides for rule by a president for a term of seven years and a National Assembly elected by universal suffrage.

Colors: *navy blue, whitish gray, vermilion, orange, pale blue, pale pink, gold*

The Moon Is Not Made of Green Cheese

The moon's surface was fine and powdery, and it adhered like powdered charcoal to the soles and sides of the astronauts' boots. They planted a three-by-five-foot American flag on the moon. It was stiffened with thin wire, so that to earthbound viewers it appeared to be flying on the windless lunar surface.

Lunar gravity makes everything weigh only one sixth of what it weighs on Earth. Despite their heavy space suits and backpacks, the astronauts felt light, and they bounded about easily like kangaroos. "Sometimes it takes about two or three paces to make sure that you've got your feet underneath you," observed Aldrin.

While the astronauts were busily hopping on the moon, a telephone call interrupted their gamboling. President Nixon was on the line from his office in the White House, to congratulate them. "The heavens have become a part of man's world," said the President, "and as you talk to us from the Sea of Tranquility, it requires us to redouble our efforts to bring peace and tranquillity to Earth."

Armstrong thanked the President in the name of men of peace of all nations.

"Resolved: that the flag of the United States be thirteen stripes, alternate red and white, that the union be thirteen stars, white in a blue field, representing a new constellation."

This resolution was adopted, on June 14, 1777, by the Second Continental Congress at Philadelphia.

Congress gave no hint as to the designer of the flag, no instruction as to the arrangement of the stars, and no information on the flag's appropriate use. As for Betsy Ross, no direct documentary evidence has been found to link her with the flag adopted by the Continental Congress.

Colors: light blue, steel blue, azure, ocher, black, vermilion, white, dark gray

Moon Chores

The moon is not exactly a good piece of real estate. It has no air and water, and its temperature ranges from about 243 degrees above Fahrenheit in the sunlight of the lunar midday to about 279 degrees below zero during the lunar night. In such a rigorous climate, the space suit with its portable life-support system is a vital necessity.

Each astronaut had a checklist of duties printed on one sleeve of his space suit. Their first assignment was to capture the "solar wind," the outflowing gases from the sun, such as argon, krypton, zenon, neon, and helium. Armstrong and Aldrin set up a sheet of aluminum foil about a foot wide and five feet long. The foil was placed so that it faced the solar rays, to which it was exposed for seventy-seven minutes. It was then packed, to be brought back to Earth. Particles of the solar-wind gases adhering to the foil could be measured and analyzed by mass spectrometers and other instruments.

Another experiment was setting up and leaving on the moon a seismometer, which would report to Earth any tremors caused by falling meteorites or volcanic eruptions. The reports would provide clues to the composition of the moon.

A two-foot-square laser reflector was also left on the moon. It was designed to reflect a beam of light directly back to Earth. By measuring the travel time of the laser beam to the moon and back, scientists could accurately measure the distance between Earth and moon and so detect subtle changes in distance. The reflector also was used to study the nature of Earth's irregular rotation, the motion of the moon, and the relative motions of Earth and moon.

India occupies the larger part of the subcontinent of India. It is one third the size of the United States. Its climate varies from the tropical heart of the south to the arctic cold of the Himalayas. Its principal food products are rice, millet, wheat, barley, coffee, spices, tea, and cashew nuts. Other important products are cotton, copra, jute, linseed, and rubber.

India's first nuclear power plant was built with United States help.

Population: 555 million. Area: 1,262,000 square miles. Capital: New Delhi.

Colors: light and dark sepia

Moon Rocks

Top priority went to collecting rock and soil samples from the moon. The astronauts used scoops and tongs with long handles, for it was not easy to bend down in the stiff space suits.

As soon as he got out of the *Eagle,* Armstrong began to collect rocks. He put them in his pant's-leg pocket to ensure having at least some lunar material if something suddenly went wrong and the astronauts had to abort the mission.

At the end of the journey Armstrong reported that he alone had collected fifty pounds of moon material.

The samples were put in sealed aluminum boxes for the trip back to Earth. The job well done, the two men climbed back into the lunar module and through the hatch threw out items no longer needed–cameras, the TV unit, hand tools such as tongs and hammer, the environmental-control-system canister and bracket, urine bags, and the two portable life-support systems. The refuse remained as a souvenir of man's first visit to the moon.

It was 1:11 A.M. Eastern Daylight Time on July 21 when they closed the hatch. Armstrong and Aldrin did some "housekeeping chores," and at 4:25 A.M. they curled up on the floor of the *Eagle* and went to sleep. Man's first sojourn on the moon had lasted twenty-one hours and thirty-seven minutes, some two hours of which were spent exploring on its surface.

In Fujeira, sheep and goats are the main domestic animals. Cows and oxen bear on their shoulders a hump similar to the hump on the so-called Brahma bull, and asses here are more often employed for riding than for carrying loads. The Fujeiran dog differs very little from a jackal in appearance; its muzzle is pointed, its color sandy, and its long and bushy tail curls over the back. Another kind of dog, similar to the greyhound and evidently of exotic origin, is often reared in Fujeira and is in great favor as a harrier.

Colors: navy blue, sea green, wisteria blue, red plum, vermilion, chrome yellow, white, black

Back to Earth

The lunar module is divided into two stages, the descent stage and the ascent stage. When the astronauts leave the moon's surface, they use the descent stage as a launching base and navigate in the ascent stage to the rendezvous with the command module. The descent stage remains on the moon as a souvenir of man's visit there. When the ascent stage is linked up with the command module, the astronauts climb into the CM and leave the ascent stage to become a piece of junk circulating in lunar orbit.

This schedule worked exactly as planned. The *Eagle*'s ascent engine burned for more than seven minutes, providing the velocity for a vertical rise toward a rendezvous with *Columbia.* The *Eagle* met *Columbia* and docked. Armstrong and Aldrin transferred through the tunnel to *Columbia,* joined Collins, closed the hatches, and jettisoned the *Eagle.*

Guinea is in western Africa. It is bordered by the Atlantic on the west, by Portuguese Guinea, Senegal, and Mali on the north, by the Ivory Coast on the east, and by Liberia and Sierra Leone on the south. Guinea and three of its neighbors—Senegal, Mali, and Mauritania—formed the Senegalese River States Organization to promote economic cooperation.

Colors: navy blue, ocher, whitish gray, steel blue, white, gold, red, blue

Peace for All Mankind

The lunar-module launch pad remains on the surface with a plaque attached that reads:

"Here men from the planet Earth first set foot upon the moon. July 1969 A.D. We came in peace for all mankind."

A touching detail: Neil Armstrong had taken with him on that historic journey to the moon medals with the bas-relief portraits of Gagarin and Komarov, the two Russian space pioneers who had lost their lives. When he visited Moscow, he gave the medals to their widows.

The independent Republic of Liberia lies on the southern coast of the West African bulge. It is bordered on the northwest by Sierra Leone, on the north by Guinea, and on the east by the Ivory Coast. Its southern border is a 350-mile coastline on the Atlantic Ocean.

Liberia was founded in 1822, when a settlement was established at Monrovia by Negro freedmen from the United States sponsored by the American Colonization Society. It was declared a republic on July 26, 1847. Its constitution is modeled on that of the United States.

Liberia's main products are citrus fruits, vegetables, tobacco, and livestock. Its industry is based on food processing, textiles, cement, and oil refining. Diamond mining is becoming increasingly important.

Population: 1.2 million. Area: 43,000 square miles. Capital: Monrovia.

Colors: violet-blue, bright red, white

Splashdown!

On the way back to earth, the command module began to heat up as it met air resistance, and the temperature of the heat shield rose to approximately 5,000 degrees Fahrenheit.

On July 24, the capsule splashed into the Pacific Ocean 825 miles southwest of Hawaii and 11 miles from the waiting carrier U.S.S. *Hornet.* A recovery frogman swam to the bobbing command module and passed three biological isolation suits with plastic-visored masks into the opened hatch to the astronauts. This was the first step in a series of rigid quarantine safety measures. Although most scientists felt the possibility remote that any dangerous microorganisms existed on the moon, the quarantine was necessary to reduce to an absolute minimum the chance of epidemics occurring here on Earth where life might lack natural defenses against alien organisms.

As soon as the astronauts emerged in their germproof suits,

they were thoroughly washed with disinfectant, after which they in turn washed the frogman in case he might have become contaminated in the process. The astronauts were then taken by helicopter to the *Hornet* and placed in a quarantine van. The van was unloaded from the *Hornet* in Hawaii and then flown to the Lunar Receiving Laboratory near Houston.

In anticipation of the termination of British control in the Trucial States, the sheikdoms in 1971 were considering a federation with neighboring arab States. Such a union, they believe, could make its weight felt in world councils as well as in the rest of the Arab world.

Colors: royal blue, white, crimson, grass green

Biological Tests

Lunar samples remained under quarantine forty-five to fifty days while NASA scientists and technicians photographed, weighed, catalogued, chipped, and even burned the rocks. Particles were tested on living cells, including those taken from fish and on human cancer. Other particles were tried out on algae, sunflowers, and pine seedlings to see the possible effect of exposure to the alien matter. Pulverized lunar rocks were fed, by injection, into Japanese quail, oysters, white mice, and cockroaches. The cockroaches were chosen because they are one of the hardiest insects known to man, having survived as a distinct genus for millions of years.

All organisms were bred and raised in a germfree environment. The mice, for example, were born by Caesarian operation in sterile surgery and raised in germfree conditions. If no lunar bugs developed after the quarantine period, the samples of lunar rocks would be sent to outside universities for further investigation.

Wherever the astronauts visited they gave their hosts a lunar rock as a souvenir. Visiting Moscow, Armstrong gave a moon rock to Alexei Kosygin, chairman of the USSR Council of Ministers, on behalf of the President of the United States and the American people. Kosygin stressed that he was particularly glad to meet Armstrong, "a man who is a symbol of progress and whose courage made such a gift possible."

You can see one moon rock exhibited in the public lobby of the United Nations Assembly Building in New York City.

Qatar is a hot and arid lowland, with an average rainfall of less than four inches. Its people are mostly Arab in origin, and their religion is of the Sunni sect of Islam.

Colors: *light and dark lavender, white, light turquoise, cream, grayish beige, dark grayish olive*

Lunokhod I

Unlike the United States, the Soviet Union has not undertaken a manned space effort in the study of the moon. The earliest schemes, starting soon after the 1957 launching of Sputnik, were for a remote-controlled moon rover, baptized Lunokhod, whose name comes from the Russian word meaning "moon rover."

Lunokhod I was delivered to the moon by Luna XVII on November 17, 1970. For almost a year, Lunokhod explored the lunar surface, testing the moon's soil, and transmitting panoramic television pictures. Lunokhod I also carried a French-built laser-reflector to enable scientists using laser beams to measure precisely the Earth-moon distance. In the first six months, the 1.663-pound eight-wheel vehicle, constructed mainly of magnesium alloy, had completed a topographic survey of a strip 2.2 miles long and 500 feet wide.

Lunokhod carries a solar panel which captures the sun's energy to charge the batteries that operate the vehicle.

Moon rovers are particularly needed for exploration of the "other" side of the moon, from which direct radio communications with the Earth are virtually impossible. The moon rover may return to the visible side of the moon and then transmit the stored information back to Earth.

The highest legislative authority in the Soviet Union is the Supreme Soviet, which meets infrequently. Between sessions, power resides in the Presidium of the Supreme Soviet, whose chairman is titular chief of state.

Elections for the Supreme Soviet are by universal suffrage but from a single slate of candidates approved by the party. Voters are offered a choice of voting for the candidates or of striking out their names.

Colors: white, violet, indigo, black-brown

Apollo 12

For Apollo 12, the command module was named *Yankee Clipper,* and the lunar module, *Intrepid.*

The launch date was November 14, 1969, and the travelers were Navy Captains Charles Conrad, Jr., and Richard F. Gordon, Jr.; and Navy Lieutenant Alan L. Bean. The target was not the Sea of Tranquility, but the Ocean of Storms, close to the unmanned American spacecraft Surveyor 3 that soft-landed on the moon on April 19, 1967, and began transmitting remote-control color-television signals back to Earth.

Conrad and Bean spent almost eight hours on two walks on the lunar surface. They explored several craters and visited the site of Surveyor 3. They set up an array of scientific instruments and left behind a nuclear-powered scientific observatory to transmit data to Earth. Also left behind were an American flag and the *Intrepid*'s lower section, which served as a launching platform for the return to orbit of the moon.

They brought back a total of seventy-five pounds of rock and soil, and the television camera which Surveyor 3 had left on the moon. One of these moon rocks was found to be 4.6 billion years old—the oldest rock yet discovered on the moon and the Earth.

On their way back to Earth the three astronauts held the first space-to-ground news conference.

On the moon the two astronauts had a television mishap. For forty-four minutes they transmitted live television pictures to 30 million people, and then, suddenly, the picture went black. Alan Bean had moved the camera directly into the sun, and the image tube burned.

The area of the Trucial States has been the scene of conflicts with Saudi Arabia. Much of the dispute involves a number of oases claimed by both parties. The planned withdrawal of the British, who have maintained peace in the area for centuries, will make imperative cooperation and resolution of differences among the area's sovereigns.

Colors: white, pink, lavender, black, red, sea-blue, aquamarine

Apollo 13

Those who believe that the number thirteen brings bad luck will find their superstition justified in the story of Apollo 13.

Apollo 13 was launched April 11, 1970, with Navy Captain James A. Lovell, Jr., and civilian test pilots Fred W. Haise, Jr., and John L. Swigert, Jr. Lovell and Haise were supposed to explore the moon, while Swigert remained in the command module in orbit. This time, the command module was called *Odyssey.* The name was ominous: the crew of Apollo 13 were to encounter as many troubles as their Greek namesakes.

Two days after launching (on April 13!) 205,000 miles from Earth, the spacecraft was rocked by "a pretty large bang." In less than a minute one of the service module's two oxygen tanks was empty. Apparently it had ruptured and a mixture of gas and liquid gushed out of the spacecraft. One of the main electrical circuits went dead, and the blast also affected two of the service module's three fuel cells, which produced the bulk of *Odyssey*'s vital electrical power. The moon landing was now out of the question. The only concern that remained was to save the astronauts' lives.

One of the dangers they faced was the loss of oxygen. It was not only essential for breathing, but also for reacting with hydrogen to produce power in the remaining fuel cell. Without oxygen there could be no return to Earth.

Ground controllers calculated that within ninety-one minutes, *Odyssey* would be completely dead and uninhabitable.

Cook Islands in the South Pacific are named after their discoverer, the famous explorer Captain James Cook. There are fifteen islands, or atolls, with a total area of ninety square miles. The Cooks became self-governing in 1965, but New Zealand retains responsibility for defense and foreign affairs.

The economy is mainly agricultural, producing citrus fruits, copra, and tomatoes as the chief exports.

Population: 20,000. Area: 90 square miles. Capital: Avarua.

Colors: gold, white, black, grass green, sky-blue, pink, golden yellow

Abandon Ship!

The crippled Apollo 13 was about 205,000 miles from home, still speeding toward the moon. Under the most ideal circumstances it would take days for the astronauts to return to Earth. To prevent the command module from becoming uninhabitable, the three men shut it down and crawled through the connecting tunnel to the lunar module, this time called *Aquarius*. Their only hope was to use it as a lifeboat that would take over all the vital functions of the crippled mother ship while still attached to it. Thus, the command module's oxygen would be saved for the reentry trip.

Since it would have required more power to turn around in midcourse, the astronauts continued on their way toward the moon. Taking advantage of its gravitational pull, they swung around it; then, after a successful engine burn, Apollo 13 was on course back toward Earth.

Thousands of miles from Earth, the astronauts were cold and thirsty. When the temperature was only a few degrees above freezing, they rubbed their hands and stamped their feet to keep warm. They called the lunar module the "refrigerator."

About thirty thousand miles from home, the astronauts

climbed back into the command module and jettisoned the lunar module, taking photographs of their lifeboat as it disappeared, with these words from Mission Control: "Farewell, *Aquarius,* and we thank you." After six days of agony for the astronauts on their space "odyssey," the command module splashed down in a perfect landing in the South Pacific.

The whole world anxiously followed the fate of the American astronauts. Pope Paul said, "We share the universal trepidation for the fate of these heroes." In Jerusalem, Orthodox Jews said special prayers at the Wailing Wall. In India, more than a hundred pilgrims held services. Conforming to the agreement concluded at the United Nations, thirteen nations, including the Soviet Union, offered ships or planes to help in the rescue operation if needed. In this hour of danger the astronauts belonged to the world.

After splashdown, President Nixon lighted up a victory cigar, then declared a national day of thanksgiving and prayer. Later, Captain Lovell told Congress that the explosion in the oxygen tank was felt as a "small shudder," rather than heard, since there is no sound in the void of space. According to the postflight investigation, a short circuit in the thermostatic mechanism controlling the flow of liquid oxygen in one of the command ship's two oxygen tanks had caused the fire and explosion.

Most of Mongolia is a steppe, or plateau, surrounded by mountains from northwest to northeast, and by the Gobi Desert in the south. Much of Mongolia's economy is animal husbandry—sheep, goats, camels, yaks, and horses—and processing of livestock products. Almost all of Mongolia's foreign trade is with the socialist countries of Europe and Asia. A mutual assistance pact was signed with the USSR in 1966.

Colors: vivid purple, peacock blue, white, red, black, lavender-pink, medium blue

Apollo 14

The Apollo 14 mission began January 31, 1971, with the lift-off at Kennedy Space Center, destination Fra Mauro on the moon. On board were Alan B. Shepard, Jr., now a captain, the first American ever to fly into space, Navy Commander Edgar D. Mitchell, and Air Force Major Stuart A. Roosa, a redhead, sometimes called Rojo, the Spanish word for "red." Since *Charlie Brown* and *Snoopy,* the astronauts had become serious, and the command service module was baptized *Kitty Hawk,* after the locality in North Carolina where the Wright brothers flew their first airplane on December 17, 1903. The lunar module was given the name of *Antares* after a large red star in the constellation of Scorpio.

The mission was planned to provide scientists with a great deal of information about the moon. The astronauts had to set up new lunar testing equipment, bring back a heavy load of geological specimens, and be more active on the lunar surface than previous moon explorers. The *Antares* landed in the Fra Mauro Formation, a rugged area of several layers of rock and debris, samples of which may help unravel the mystery of the moon's origin.

Oil was discovered in Qatar in 1939, and production began on a large scale after World War Two. Now the income from oil adds about $70 million annually to Qatar's economy, and the proceeds have been used for road construction, a seawater conversion plant, and other needed developments.

Colors: navy blue, steel gray, ocher, white

APOLLO 14
Shepard
Roosa
Mitchell
31.I.1971
h 21 03' 20'' GMT
DECOLARE
05.II.1971
h 9 18' 15'' GMT
ASELENIZARE
09.II.1971
h 21 04' GMT
AMERIZARE
APOLLO 14
1971
ALAN B.
SHEPARD
APOLLO 14
posta aeriana
3,30 L
POSTA
ROMANA
APOLLO 14
posta aeriana
3,30 L
POSTA
ROMANA
APOLLO 14
posta aeriana
3,30 L
POSTA
ROMANA
APOLLO 14
posta aeriana
3,30 L
POSTA
ROMANA
APOLLO 14
1971
EDGAR D.
MITCHELL
APOLLO 14
1971
STUART
A. ROOSA

Football-Sized Moon Rocks

This time the astronauts had with them a two-wheeled, two-legged cart to carry tools and lunar-sample payloads. It proved to be a useful vehicle, though several times they had to carry it when it seemed to be actually slowing them down.

Shepard and Mitchell retrieved rock samples from the 18-degree slope of Cone Crater; but the difficulties in bending the knees and ankles in the space suit tended to tire them.

The astronauts made two excursions on the moon with about four hours of rest in between. They spent a record total of nine hours, twenty-five minutes in extravehicular activity, having traveled nine thousand feet from *Antares*. They dug deep into the lunar surface and brought back approximately ninety-six pounds of lunar material, including two football-sized rocks and many small glasslike pebbles.

From instruments deployed on the lunar surface by the Apollo 14 mission—such as the passive seismic experiment package—scientists have discovered that quakes occur deep within the moon, indicating that it may contain a core of molten material. "It turns out that the moon is not as dead as everyone expected," said seismologist Gary Latham of the Lamont-Doherty Geological Observatory.

In 1969, President Nixon visited Rumania. It was the first such visit by an American President in twenty-four years. Rumania has close economic ties with Western Europe as well as with the socialist countries, and was the first of the Eastern-bloc nations to establish diplomatic relations with West Germany.

Colors: deep blue, terra cotta, light gray, black, white, silver

Lunar Golf Balls

On February 6, 1971, Fra Mauro base became the first lunar golf course. On that day Captain Alan B. Shepard, Jr., with the permission of Mission Control Center, pulled a golf ball out of his pocket, put it down in the lunar dust and, using the handle of a sample collecting device with a genuine number-six iron on the bottom, drove the ball "a couple of hundred yards." He made three attempts, and the third drive went about four hundred yards, with Shepard adding, "This was not bad for a six iron." Golf balls made history.

The Republic of Togo on the Gulf of Guinea in West Africa, a former German colony, has been a French mandate, then a French trusteeship, and finally, in 1960, became an independent republic. Though French is the official language, most Togolese speak Ewe, Twi, or Hausa, or one of forty-four local dialects.

Togo's principal products include coffee, cocoa, yams, corn, millet, and other agricultural produce; phosphates are abundant and mining of iron ore is beoming of major importance.

Population: 1.9 million. Area: 22,000 square miles. Capital: Lomé.

Colors: sea green and cream

Mission to Mars

While the preparation for landing on the moon was in progress, both Russia and the United States looked for new areas to explore in the universe. Since 1960 the Russians have sent a number of successful probes toward Mars, and their radios have transmitted information about it. The United States Mariner 4 took the first surface pictures of Mars as it passed the planet in 1965. In mid-1969 Mariners 6 and 7 took excellent photographs of Mars from less than 2,000 miles away.

In May of 1971 the United States and the Soviet Union each launched two Mars probes. Mariner 8 did not reach a sufficient velocity to escape the Earth's atmosphere. Mariner 9, after a trip of over five months and 248 million miles, reached Mars on schedule early in November and swung into an oval orbit around the red planet, coming as close as 800 miles to its surface. This was the first time a man-made satellite had orbited another planet. Although the first photographs were blurred because of a dust storm, it was learned that the planet has craters, a huge bulge at its equator, and wide irregularities in its gravitational field.

There is no state religion in Hungary. About two thirds of

the population are Roman Catholics; most of the remainder are Presbyterian and Lutheran. All religions are tolerated. Though the word "Magyar" is today synonymous with Hungarian, the original Magyars were the fierce horsemen from beyond the Urals who conquered and settled in Hungary a thousand years ago.

Colors: cornflower blue, black, white, khaki, yellow

Luna XVI

In September of 1970, the Soviets launched Luna XVI, an unmanned vehicle that landed on the moon. It was controlled entirely by a computer program fed from Earth. On a signal, an electric-powered drilling device, capable of moving horizontally and vertically, reached out like a mechanical hand, burrowed about fourteen inches into the lunar soil, and brought up a core sample, which was packed into a container, hermetically sealed, and then safely stowed away in the upper half of the spacecraft.

When the Earth and moon were in proper alignment, Luna XVI's upper-stage rocket was ignited, hurtling the vehicle back toward Earth. The four-legged lower stage remained on the moon and continued to take temperature and radiation readings.

Twelve days after it had been sent aloft, Luna XVI landed in central Asia. The lunar sample was quickly flown to Moscow, where it was turned over for analysis to the Soviet Academy of Sciences.

The Sea of Fertility, where Luna XVI landed, is thought to be one of the oldest lunar maria, and the soil that was retrieved from below its surface yielded important clues to the early history of the moon.

Mongolia is the most thinly populated country in the world. Tibetan Buddhism is the dominant religion, but only a small number of lamas remain to support the old culture, which is rapidly giving way to the new way of life. Education, industrialization, and growing contact with the outside world are bringing vast changes to Mongolian life.

Colors: emerald green, fiery orange, white, ocher, peacock blue, lavender, aquamarine, olive-gray

Mission to Venus

An unmanned Soviet spacecraft, Venera 7, reached Venus on December 15, 1970. It survived in the murky atmosphere of the planet and continued to function after landing, remaining in operation for twenty minutes in an atmospheric pressure ninety times as high as that on Earth and in a temperature of 900 degrees Fahrenheit.

Venera 7's radio transmissions carried the first scientific data to be transmitted to Earth from the surface of another planet. The fact that the instruments continued to work for as long as they did in the intense heat was a testimonial to the increasing sophistication of the Soviet space program.

Venera 7 was the fourth successful Soviet mission to Venus. Although Veneras 4, 5, and 6 were on target, deployed parachutes, and transmitted data after entering the planet's atmosphere, none reached the surface in operating condition; they were either crushed, or overheated while floating down through the dense, boiling atmosphere.

In 1962, the United States sent Mariner 2 within 21,000 miles of Venus.

The Magyars embraced Christianity in the tenth century and ruled Hungary for about four hundred years. The country suffered repeated Turkish invasions in the fifteenth to seventeenth centuries. In 1686 the Austrians replaced the Turks as conquerors, and Hungary became part of the Hapsburg Empire. In World War Two, Hungary joined Germany. Russian troops occupied the country in 1944, and a Soviet-type constitution was adopted in 1949.

Colors: cornflower blue, white, black

Orbiting Shish Kebab

"Shish Kebab" is the pet name for the unmanned space station Salyut ("Salute") which the Russians launched on April 19, 1971, on the tenth anniversary of man's first space flight by Colonel Yuri A. Gagarin. Seven years after his historic trip, Gagarin was killed in the crash of a training plane.

The Salyut is nicknamed Shish Kebab because a number of manned spacecraft will be attached to it. In fact, three days after its launching, a three-man crew in Soyuz X ("Union") joined it in a rendezvous and docked successfully for five and a half hours.

Salyut will make it possible to engage in an all-around investigation of the globe and space near Earth in the interest of meteorology, geophysics, oceanology, and other branches of science. The president of the Soviet Academy of Science, Matislav V. Keldysh, even mentioned plans to gather solar energy at such stations and beam it to Earth for conversion to electrical power.

The Republic of Ghana is composed of former British colonies and protectorates: Gold Coast, Northern Territories,

Ashanti kingdom. Ghana is bordered by Upper Volta on the north, by Togo on the east, by the Ivory Coast on the west, and by the Gulf of Guinea on the south. Ghanaians are mainly Sudanese in origin; the principal groups include Akans, Ashanti, and Guans.

Ghana is the world's leading cocoa producer.

Population: 9 million. Area: 92,000 square miles. Capital: Accra.

Colors: moss green, red, terra cotta, grayish lavender

"Bridge to the Future"

The United States is preparing a 48-foot-long orbiting workshop, Skylab, that is expected to catapult astronauts into space voyages lasting months or even years. Skylab, the "Bridge to the Future," will be a giant Earth-orbiting laboratory which three crews of astronauts will inhabit over a period of eight months. Skylab will have water and breathing gases (nitrogen and oxygen) lasting three men for eight months and a sufficient supply of food.

The Skylab astronauts will work ten-hour days, sleeping and eating at about the same time as the people in the United States. After the evening meal they may relax with movies, games, and books, or may be collecting their thoughts or lining up the experimental equipment for the next day.

Their living quarters will include a kitchen, food freezers, and bicycle-type exercise machines.

The astronauts will dispose of their waste by storing it in a tank that would have contained the rocket propellant in the ascent stage. Samples of their urine will be frozen and returned to Earth for analysis every day.

Three missions are planned for Skylab. The first one, in which three men are to live in the workshop for twenty-eight days, would be to study the physiological and psychological aspect of flight. The second mission, planned for fifty-six days, will be devoted primarily to operating the telescopes. And the third crew, during its fifty-six days, will specialize in experiments to survey Earth resources with highly sensitive cameras.

The three-man workshop is to be launched early in 1973, according to plans. A full-fledged general-purpose laboratory is to be operational in the early 1980's.

Many Americans receive awards and prizes each year for their contributions to the advancement of the space program. For instance, the American Institute of Aeronautics and Astronautics conferred these awards in 1970: Goddard Award for contributions in the engineering science of propulsion or energy conversion (Gerhard Naumann); Robert M. Lasey Award for contribution to the science of meteorology (Newton A. Lieurance); Space Science Award for investigation of the physics of atmosphere of celestial bodies (Carl McIlwain); Sylvanus Albert Reed Award for contribution to aeronautical engineering design (Richard T. Whitcomb); Haley Astronautics Award to test personnel who undergo personal risk in the advancement of space flight (William Anders, James A. Lovell, Jr., and Frank Borman): Octoave Chanute Award for contribution by a pilot to the aerospace sciences (Gerauld Gentry).

Colors: strong purple and white

The Hatch Was Opened—Three Men Were Dead

Three Soviet cosmonauts, Lieutenant Colonel Georgi T. Dobrovolsky, Flight Engineer Vladislav N. Volkov, and Test Engineer Viktor I. Patsayev, on June 7, 1971, maneuvered their craft, Soyuz XI, to a rendezvous with space station Salyut and then boarded the station's spacious crew quarters to inaugurate the first manned scientific workshop in orbit around the Earth.

The achievement gave the Soviet Union a two-year head start over the United States' plan to place a three-man workshop, the Skylab, into Earth orbit in 1973.

The three cosmonauts spent nearly twenty-four days in space and set an endurance record.

On June 29, 1971, the crew completed the flight program in full and were directed to transfer to Soyuz XI and return to Earth. The unlinking operation passed without a hitch, and on June 30, Soyuz XI landed in the preset area. The recovery group, upon opening the hatch, found the three cosmonauts in their seats—dead.

The official report from the Soviet Union concluded: ". . . thirty minutes before landing, there occurred a rapid drop of pressure within the descent vehicle which led to the sudden death of the astronauts . . . the drop in pressure resulted from a loss of the ship's sealing."

The loss of oxygen quickly rendered the three men unconscious and brought about a rapid, painless death.

The heart of mankind went out to the Russian people in their hour of tragedy. The American astronauts sent a cable to Brigadier General Andrei G. Kuznetsov, which said: ". . . We mourn with all the Russian people the death of three of their brave countrymen." Pope Paul VI said, "We express our grief for this unexpected and tragic ending to their endeavour, which has aroused so much admiration. . . ." President Nixon wrote his own longhand condolence to Maria and Natasha, the daughters of cosmonaut Dobrovolsky: ". . . I also have two daughters," the President wrote, "and their hearts go out to you as do Mrs. Nixon's and mine."

Leonid I. Brezhnev, Communist party leader, wept at the funeral.

The tragedy didn't deter the Soviet space effort. On September 2, 1971, Luna XVIII, a new unmanned space vehicle, was launched to explore the moon.

The Hungarian language, Magyar, is one of the Finnic-Ugric languages. It is written in Roman alphabet.

Colors: turquoise, white, black, silver, medium brown, flesh color

Post Office on the Moon

On July 26, 1971, three American astronauts, David R. Scott, Air Force Lieutenant Colonel James B. Irwin, and Air Force Major Alfred M. Worden soared toward the moon on board Apollo 15. It was called the smoothest and most nearly flawless start, and it was the fourth American moon-landing venture in two years. Scott and Irwin rode the lunar module *Falcon,* while Worden was at the control of the command ship *Endeavor,* which remained in lunar orbit.

This time the astronauts had on board the *Falcon* a jeeplike four-wheeled lunar rover vehicle (LRV), baptized *Rover.* The astronauts were to use it on the moon's surface.

On July 31, the two astronauts landed on the surface of the moon and unpacked the *Rover* and tested it. There was the possible danger that in the low gravity of the moon the LRV would leave the ground when it hit a bump. The astronauts buckled their seat belts, but even so there were times when they had the feeling that they were riding on a bucking bronco.

Scott and Irwin spent a total of sixty-seven man-hours on the moon and made three trips, returning between trips to the *Falcon* for a rest. They covered more than seventeen miles in the *Rover,* traveling to the very edge of a winding quarter-mile-deep gorge called Hadley Rille. They sent back excellent color-television pictures, took fourteen hundred still photographs, and brought back 180 pounds of rock, among them crystalline fragments called Genesis rocks, believed to be remnants of the primordial lunar crust of some 4.6 billion years ago. The Genesis rocks are common on earth, but that they can be found on the moon gives food for thought. It will take scientists up to ten years to evaluate these rocks and shed some light on the origin of the moon and Earth.

The two men left on the moon a touching memorial to the fourteen American and Soviet astronauts who have died since manned space exploration began a decade ago. It was a metal plaque with the names of the dead space voyagers and a small figurine symbolizing "the fallen astronauts."

Colonel Scott opened the first lunar post office, saying, "To show that our good postal service delivers anyplace in the universe, I have the pleasant task of canceling here on the moon the first stamp of a new issue dedicated to commemorate the United States achievement in space. I'm so proud to have the opportunity here to play postman."

Colors: blue-black, Wedgwood blue, white, medium gray, yellow, red

Space Technology—Its Practical Applications

Since the first man-made satellites, Sputnik I and Explorer I, thousands of satellites have been sent into orbit around the Earth and to penetrate farther and farther into our solar system and beyond. By July 26, 1971, there were about 2,500 objects circulating in space, and close to 3,000 had already decayed. In a short period of ten years, space science and space technology have created a vast potential not only for the better understanding of the universe but also for practical benefits for the peoples of the world. Space research has become an important tool of modern science and technology. It affects man's daily life and will affect it even more in the future.

Satellites will supply data on air and water pollution and on other hazardous developments threatening the environment. Constant monitoring of the polar icecaps should make it possible to make accurate predictions of the runoffs in the spring and identify potential flood areas.

Through satellite weather forecasting and storm warnings,

countless thousands of lives have already been saved and property damage has been mimimized. International calls via satellite reached a volume of ten million in 1970.

Every year about a thousand technological innovations generated in the space program find their way into our earthly industries, where they lead to better kitchen appliances (teflon) and farm equipment, improved sewing machines and radios, better ships and airplanes, even advanced computers and communications equipment, and life-saving medical instruments.

Under the stimulus of space exploration, medical science has made progress in microbiology, nutrition, mental health, and cardiology. Medical communication science, which applies such techniques as telediagnosis and telesurveillance, has developed in a spectacular way.

In 1972 the first Earth-resources-technology satellite will be launched. Predictions of benefits range into the billions of dollars for agriculture, hydrology, ocean transportation, fishing, and mineral resources.

All these are "spinoffs," or side benefits, derived from the main stream of space effort.

Space exploration is becoming increasingly complex, and Soviet and American leaders have realized that collaboration among nations has become a necessity. Such collaboration might be realized under the flag of the United Nations.

The Commonwealth of Australia is an island subcontinent five times the size of Alaska. It is situated southeast of Asia proper and south of the islands of Indonesia. Australia is part of the British Commonwealth of Nations and Queen Elizabeth is its sovereign. It is, however, an independent state with its own legislature and prime minister.

Wool is Australia's greatest industry, with an annual clip of 1.9 billion pounds.

Population: 12.5 million. Area: 2,968,000 square miles. Capital: Canberra.

Colors: ultramarine blue, sky blue, black, white

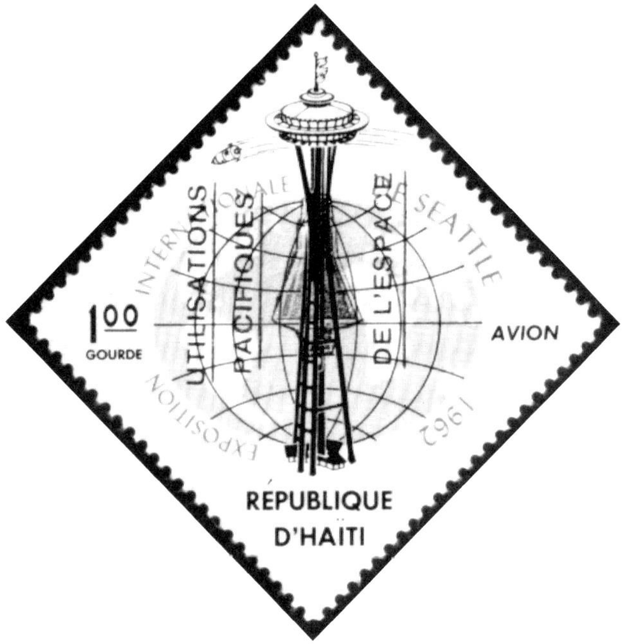

Medical Miracles Through Space Research

The medical benefits resulting from space exploration verge on the miraculous. There is a wheelless wheelchair for crippled children that can clear curbs and climb stairs. It was developed from a walking vehicle built for NASA for exploring the moon's surface. The wheelless wheelchair can serve children who can use neither arms nor legs and must steer the vehicle with their hands.

A modified version of an astronaut's helmet can be used to measure the oxygen consumption of a person in action. Old-fashioned respirometers required the patient to wear a nose clip and inhale and exhale through a mouthpiece, an uncomfortable and unreliable procedure.

A highly sensitive instrument devised for measuring micrometeorite strikes on spacecrafts was modified to record the tiny muscle tremors associated with the onset of Parkinson disease, a shaking paralysis.

A new lightweight, durable alloy was developed in NASA's laboratory near Cleveland, Ohio, whose use has now been extended to artificial joints.

Sight switches assist a patient who cannot move his hands or legs. This is a miraculous device mounted at each side of a pair of eyeglasses. The device can detect the difference between the reflection of the white of the eyes and the darker pupils, and activate an electric switch. By remote-control operation a patient can turn pages of a book, tune in on radio or television programs, or operate his wheelchair solely with his eyes. The device was developed for NASA as a potential aid to astronauts in situations where high G-forces (G for gravity) might make them unable to move their arms or legs.

These are only a few of the medical benefits derived from space research.

Haiti was discovered by Columbus in 1492. In 1677 it became a French colony. In 1804, following a rebellion begun in 1793 by Toussaint L'Ouverture, a former slave, Haiti attained its independence. It has been a republic since 1820, although François Duvalier, who became president in 1957, established a military dictatorship.

Roman Catholicism is the state religion, but voodoo, which reaches back to the African past of the Haitian people, is also widely practiced, and often some of the voodoo rites are incorporated in a Roman Catholic ceremony.

Colors: black, white, light gray, vermilion

International Cooperation

The United Nations Committee on Outer Space was a topic of discussion at the General Assembly in 1958; in 1962 the Committee on the Peaceful Uses of Outer Space had been established. To this end it promotes international cooperation in the scientific and technical aspects of space activities and helps to develop international law to govern these activities.

A treaty which bans nuclear weapons from outer space and prohibits claims of national sovereignty in space was signed in Moscow, London, and Washington in January, 1967. So the moon belongs to everyone.

In the fall of 1967 the General Assembly unanimously approved the Agreement on the Rescue of Astronauts, the Return of Astronauts and the Return of Objects Launched into Outer Space.

Indeed, it was an uplifting moment when Apollo 13 was in trouble and thirteen nations, including the Soviet Union, offered airplanes and ships to save the astronauts' lives.

The airmail stamp shown was issued by the United Nations. The ray which pierces deep into the universe points to the future, when outer space will be used for swifter communication between peoples and for the enlightenment of man.

Colors: blue, dark blue, black, white

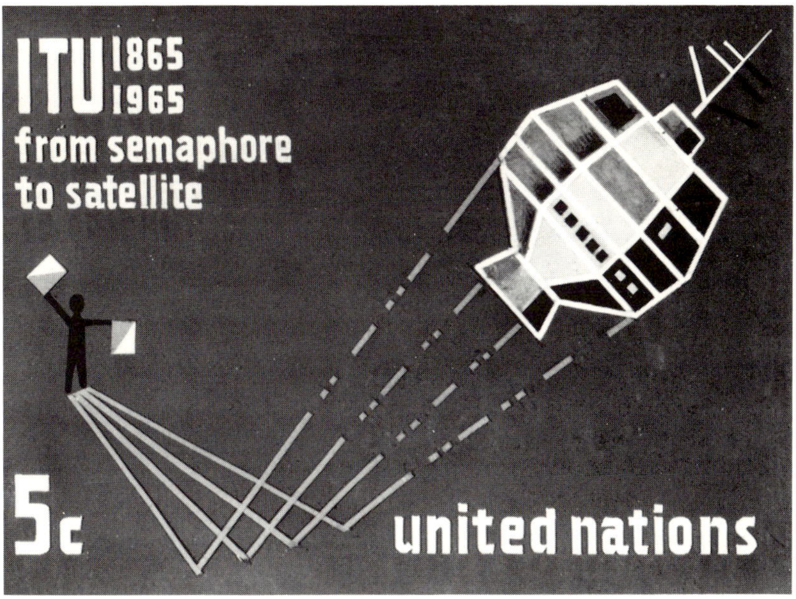

The UN in Action

In 1968 the United Nations Conference on the Exploration and Peaceful Uses of Outer Space was held in Vienna. Seventy-eight countries, three intergovernmental space organizations, and eight specialized agencies of the United Nations participated in the conference. Among them were the World Health Organization, the World Meteorological Organization, the International Civil Aviation Organization, and the International Telecommunication Union, the centenary of which was celebrated in 1965 by many countries by issuing stamps.

One hundred eighty-eight papers were presented at the

United Nations conference. In these, delegates examined the practical benefits being derived from space exploration. They reported about their countries' achievements in space and envisaged the opportunities available for international cooperation.

At the conference in Vienna, delegates asked the question: "Can one afford to undertake space research?" Dr. Vikram A. Sarabhai of India replied: "Can anyone afford to ignore the application of space research?"

This stamp was issued by the UN to commemorate the centenary of the International Telecommunication Union, 1865–1965.

Colors: aquamarine, orange, blue, purple

Epilogue

It is most encouraging that there have been, and continue to be, programs of cooperation, at all levels, in outer space research and its practical application. . . . The success of these efforts encourages us to believe that we can now press forward to greater levels of cooperation to harness the practical benefits that can accrue to man from the vast technological and scientific advances through space exploration.

U Thant
Secretary General
of the United Nations

This stamp was issued by the United Nations. It symbolizes the peaceful uses of outer space.

Color: maroon

Index

A

Accidents in spacecraft, 71, 72, 109, 126
Activities, extravehicular, 64
Airlock, 64
Aldrin, Edwin, 87
Alouette 1, 58
American astronauts, 42
Anders, William A., 75
Animals in space, 37, 39
Apollo 7, 74
Apollo 8, 75
Apollo 9, 80
Apollo 10, 82, 85
Apollo 11, 87, 89
Apollo 12, 107
Apollo 13, 109, 111
Apollo 14, 115
Apollo 15, 128
Apollo program, 65
Ariel 1, 57
Aristotle, 14
Armstrong, Neil, 50, 70, 87, 91
Ascent stage, lunar-module, 98
Astronauts, American, 42
Astronomy, principles of dynamic, 19
Atmospheric research satellite, 58
Australia, 131

B

Bahrain, 73
Bean, Allen L., 107
Belyayev, Pavel I., 64
Biological specimens in moon orbit, 62
Biological studies of moon samples, 103
Borman, Frank, 75
Brahe, Tycho, 14
Bykovsky, Valery F., 50

C

Cameroon, Federal Republic of, 55
Canada, 58
  satellites of, 58
Cape Canaveral. *See* Cape Kennedy.
Cape Kennedy, 25
Carpenter, M. Scott, 42, 51, 65
Cernan, Eugene A., 70, 83
Chaffee, Roger B., 71
Climate, lunar, 94
Collins, Michael, 87
Communications satellites, 54
Conrad, Charles Jr., 67, 107
Contributions of space technology, 131
Cook Islands, 110
Cooper, L. Gordon, 42, 51, 65, 67
Cooperation, international, 58, 131
Copernicus, 13, 14
Cosmic Rockets. *See* Lunik Series.
Cosmos I, 36
Cosmos satellites, 36
Cunningham, R. Walter, 74
Czechoslovakia, 37

D

Descent stage, lunar-module, 98
Discoverer program, 54
Dobrovolsky, Georgi T., 126
Docking in space, 70
Dog in space, 29, 31
Dynamic astronomy, principles of, 19

E

Earth
  photography of, 74
  shape of the, 54
Effects of space travel, 70
Eilmer, 14
Eisele, Donn F., 74
England, satellites of, 57
European Space Research Organization, 58
Experiments, medical, 70
Exploration
  international cooperation in space, 58
  lunar, 59, 61, 91, 113

and medicine, space, 131
unmanned lunar, 120
Explorer I, 32
Explorer 35, 61
Explorer Series, 54
Extravehicular Activities (EVA's), 66
Extravehicular Activities suit, 67

F

Feoktistov, Konstantin P., 52
Flight, manned lunar, 79
Flying machines, human, 13, 14
Forecasting, weather, 35
Fra Mauro, 113
France, Republic of, 21, 22
satellites of, 57
Freedom 7, 43, 44
Friendship 7, 47
Fuels. *See* propellants.
Fujeira, 62, 97
Functions of satellites, 130

G

Gagarin, Yuri, 41, 123
Galilei, Galileo, 17
Gemini IV, 67
Gemini V, 67
Gemini VIII, 70
Gemini IX, 70
Gemini Series, 65, 66, 70
Genesis rocks, 129
German Democratic Republic (East Germany), 29, 50, 70
Germany, Federal Republic of, 70
Ghana, Republic of, 123
Glenn, John H., Jr., 42, 47, 65
Global weather patterns, 74
Goddard, Robert H., 24
Golf on the moon, 117
Gordon, Richard F., Jr., 107
Gravity, 19
Grissom, Virgil I., 42, 44, 65, 66, 71
Guinea, Republic of, 71, 86, 88, 92 93

H

Haise, Fred W., Jr., 109
Haiti, Republic of, 35, 133
Hale, Edward Everett, 21
Hong Kong, 33
Human flying machines, 13, 14
Hungarian People's Republic, 27, 43, 119, 122, 127
Hydrogen, liquid, 23, 24

I

Icarus, 13
India, 95
International cooperation, 58, 131
Iran, 82
Irwin, James B., 128
Israel, 16
Italy, Republic of, 18
satellites of, 57

K

Kai Ka'us, 13
Kepler, Johannes, 15
Komarov, Vladimir M., 52, 72
Korea, Republic of, 90

L

Laika, 29, 31
Landing
lunar, 60, 89
preparation for manned lunar, 87
Laser reflector, 95
Law of motion, second, 19
Laws of planetary motion, 15
Leonov, Aleksei A., 64
Liberia, Republic of, 100
Liberty Bell 7, 44
Liquid hydrogen, 23, 24
Liquid oxygen, 23, 24
Lovell, James A., Jr., 75, 109
Luna I, 59
Luna II, 59
Luna III, 33, 59
Luna IX, 60
Luna XIII, 61
Luna XIV, 61
Luna XV, 90
Luna XVI, 120
Luna XVII, 105
Luna Series, 59. *See* also Lunik Series.
Lunar. *See also* Moon.
exploration, 59, 91, 120

flight, 79
landing, 87, 89
orbit, 75
surface, 93
Lunar module (LM), 98
Lunar-orbit rendezvous (LOR), 79, 86
Lunar rover (LRV), 128
Lunik Series, 30
Lunokhod I, 105

M

MacDivitt, James A., 66, 80
Magnetic field, study of, 36
Mali Republic, 51, 64
Manned lunar flight, 79
Manned lunar landing, preparation for, 87
Manned Spacecraft Center, 68
Mariner 2, 122
Mariner 4, 119
Mariner 8, 119
Mariner 9, 119
Mars probes, 119
Mauritania, Islamic Republic of, 57
Medical experiments, 70, 74
Medicine and space exploration, 131, 132
Méliès, Georges, 21, 22
Men
in space, 41
on the moon, 87
Mercury Project, 45, 65
Mercury-Redstone 2, 39
Meteorological satellites, 35
Mitchell, Edgar D., 113
Module, lunar, 98
Molniya communications satellite, 54
Mongolian People's Republic, 30, 41, 112
Montgolfier, Étienne, 31
Moon. *See also* Lunar.
far side of the, 34
golf on the, 117
landing on the, 60
men on the, 87
orbit, biological specimens in, 62
orbiting the, 62
photography of the, 76
probes, Soviet, 30, 33
rocks, 96, 110
seismometer on the, 95
soil samples of the, 96, 103
surface of the, 61
Moon rover, 105
Motion
laws of planetary, 15
second law of, 19
Multistage rocket, 23

N

National Aeronautics and Space Administration (NASA), 35
Newton, Sir Isaac, 19
Nicaragua, 67
Niger, Republic of, 52, 66
Nikolayev, Andrian G., 49, 50

O

Oberth, Hermann, 21
Ocean of Storms, 107
Orbit, moon, 62, 75
biological specimens in, 62
Orbit, rendezvous, lunar-, 79, 86
Outer Mongolia. *See* Mongolian People's Republic.
Outer space and the United Nations, 5, 28
Oxygen, liquid, 23, 24

P

Paraguay, Republic of, 19, 44, 76
Patsayev, Viktor I., 126
Patterns, global weather, 74
Photography
of the earth, 74
of the moon, 76
Pioneer 4, 59
Planetary motion, laws of, 15
Plato, 14
Polish People's Republic, 14, 49
Popovich, Pavel R., 49
Predictions, science-fiction, 20, 21
Preparation for manned lunar landing, 87
Principles
of dynamic astronomy, 19
of rocket propulsion, 19

Probes
  Soviet moon, 30, 33
  space, 119
  Venus, 121
Propellants, 23, 24
Propulsion, 26
  principles of rocket, 19
Proton series, 55
Ptolemy, 14

Q

Qatar, 75, 79, 104, 114
Quarantine after space flights, 101

R

Radiation, study of, 36
Radiation belts, Van Allen, 54
Ranger 7, 60
Ranger 8, 60
Ranger 9, 60
Ras al Khaima, 48
Reflector, laser, 95
Relay I, 56
Rendezvous
  lunar-orbit, 86
  in space, 85
Research satellite, atmospheric, 58
Rocket, multistage, 23
Rocket propulsion, principle of, 19
Rocket train. *See* Multistage rocket.
Rockets, 26
  and space travel, 22
  and war, 27
Rocks
  Genesis, 129
  moon, 96, 116, 128
Roosa, Stuart A., 113
Rover, moon, 105
Rumania, Socialist Republic of, 32, 40, 60, 80, 84, 116

S

Salyut, 123
Samples
  biological studies of moon, 103
  moon soil, 96
San Marco I satellite, 57
Satellites
  atmospheric research, 58
  Canadian, 58
  communications, 54
  English, 57
  French, 57
  functions of, 130
  Italian, 57
  meteorological, 35
  San Marco I, 57
  scientific, 54
  Syncom, 56
Schweickart, Russell L., 80
Schirra, Walter M., Jr., 42, 51, 65, 74
Science-fiction predictions, 20, 21
Scientific satellites, 54
Scott, David R., 70, 80, 128
Sea of Fertility, 121
Sea of Tranquility, 85, 89, 107
Second law of motion, 19
Seismometer on the moon, 95
Senegal, Republic of, 56
Shape of the Earth, 54
Shepard, Alan B., Jr., 42, 43, 65, 113, 117
Shoes, space, 82
Siemienowicz, Kazimierz, 27
Skylab, 124
Slayton, Donald K., 42, 65
Soil samples, moon, 96
Solar system, theories of, 14
Solar winds, 61, 94
Soviet moon probes, 30, 33
Soyuz I, 72
Soyuz X, 123
Soyuz XI, 126
Space
  animals in, 37, 39
  docking in, 70
  dog in, 29, 31
  men in, 41
  probes of, 119
  and the United Nations, 5, 28
  walks in, 64, 67, 70
Space exploration
  international cooperation in, 58
  and medicine, 131, 132
  and rockets, 22
Space shoes, 82
Space suit, 81
Space technology, contributions of, 131
Space vehicles, study of, 36

Spacecraft, accidents in, 71, 72, 109, 126
Specimens in moon orbit, biological, 62
Sputnik I, 28
Sputnik II, 29, 31
Sputnik IX, 37
Sputnik X, 37
Stafford, Thomas P., 70, 83
Stages, lunar-module, 98
Studies
of magnetic field, 36
medical, 74
of moon samples, 103
of radiation, 36
of space vehicles, 36
Suit
for extravehicular activities, 67
space, 81
Surface, moon, 61, 93
Surveyor 1, 61
Surveyor 3, 107
Surveyor 7, 61
Swigert, John L., Jr., 109
Syncom I, 56
Syncom II, 56
Syncom III, 56
Syncom satellites, 56

T

Technology, contributions of space, 131
Telescope, 17
Telstar, 54
Tereshkova, Valentina V., 50
Theories of the solar system, 14
Tiros I, 35
Tiros Series, 35
Titov, Gherman S., 46
Togo, Republic of, 117
Trucial States, 108
Tsiolkovsky, Konstantin E., 22

U

Union of Soviet Socialist Republics (U.S.S.R.), 23, 34, 38, 39, 63, 106
United Nations and outer space, the, 5, 28, 134, 135
Unmanned lunar exploration, 120

V

Van Allen, James A., 33
Van Allen Belt, 33, 54
Vanguard Series, 54
Vehicles, study of space, 36
Venera 7, 121
Venus probes, 121
Verne, Jules, 20
Vietnam, Democratic Republic of, 46, 59
Volkov, Vladislav N., 126
Voskhod II, 64, 66
Voskhod Series, 52
Vostok I, 41
Vostok II, 46
Vostok III, 49
Vostok IV, 49
Vostok V, 50
Vostok VI, 50

W

Walks in space, 64
Wars and rockets, 27
Weather
forecasting, 35
patterns, 74
Wells, H. G., 21
White, Edward H. II, 66, 71
Winds, solar, 61, 94
Worden, Alfred M., 128

Y

Yegorov, Boris G., 52
Young, John W., 66, 83

Z

Zond 5, 62
Zond 6, 63
Zond 7, 63